Lifestyles
An introduction to biology

This

Owen Bishop

Lifestyles
An introduction to biology

M

Macmillan Education

First published 1979 by
MACMILLAN EDUCATION LIMITED
Houndmills Basingstoke Hampshire RG21 2XS
and London
Associated companies in Delhi Dublin
Hong Kong Johannesburg Lagos Melbourne
New York Singapore and Tokyo

Printed in Hong Kong

British Library Cataloguing in Publication Data
Bishop, Owen Neville
Lifestyles.
1. Biology
I. Title
574   QH308.7
ISBN 0–333–23413–8

Contents

Acknowledgements

The author and publishers wish to acknowledge the following photograph sources:

Heather Angel *Cover* pp. 5 right, 16 bottom right, 17, 18 bottom left, 18 right, 19, 20 bottom, 21 top, 29 left, 29 bottom right, 46 bottom, 52, 54 bottom left, 55 left, 55 top right, 56, 57, 58 top left, 66 bottom left, 67 bottom left and right, 73, 74 top left, 75 middle left, 79, 80, 81, 82, 83 top, 84 right, 85 right, 87, 88, 89, 90, 99 bottom, 103.
Aerofilms p. 4.
Associated Press pp. 1, 2.
Peter Baker p. 30 bottom.
Barnabys Picture Library p. 37 left, 85 left.
Beecham Group Ltd. p. 68 left.
S Summerhays/Biofotos p. 99 top.
Camera Press Ltd. pp. 42 top right, 53 top, middle and bottom right, 54 top left.
J Allan Cash Ltd. pp. 3 right, 53 top left.
Ron Chapman pp. 46 top, 53 bottom left.
Bruce Coleman Ltd. pp. 5 left, 77, 78, 91 top left.
Crown Copyright, Central Office of Information p. 29 top right.
Department of Environment, Building Research Establishment p. 21 bottom.
Dolphin Showers p. 63 bottom right.
Lieselotte Evenari p. 38.
F.A.O. p. 91 right.
Forestry Commission p. 95.
Walter Gardiner-Gerrard & Co. p. 97 right.
I.C.I. pp. 54 top and bottom right, 67 top right.
India Office Library p. 42 top left.
Mansell Collection pp. 23, 26 top.
D. P. Watson, Marine Biological Laboratory, Plymouth p. 105.
N.A.S.A. p. 3 left.

N.H.P.A. pp. 34, 45.
Northern Ireland Tourist Board p. 30 top left.
Dr Clifford Butler, Nuffield Foundation p. 62 top right.
Oxoid Ltd. p. 69.
Paul Popper Ltd. p. 71.
Radio Times Hulton Picture Library pp. 26 bottom, 33.
Graham Read p. 74 bottom right.
Rentokil Ltd. pp. 55 bottom right, 63 bottom left.
Shell Photographic Service pp. 58 bottom right, 61 bottom left and right, 74 bottom left, 83 bottom.
Harry Smith Horticultural Photographic Collection pp. 18 top left, 30 top right, 94 left.
Southern Water Authority pp. 37 right, 51.
Syndication International pp. 48 right, 49 left, 91 bottom left.
Eric Taylor p. 42 bottom left.
Tesco Stores Ltd. p. 62 bottom.
Don Tindall p. 75 top right.
John Topham Picture Library p. 58 bottom left, Jane Burton p. 63 top centre left, S. C. Bisserot 84 left.
United Kingdom Atomic Energy Authority p. 54 middle left.
United States Department of Agriculture p. 75 bottom left.
C. James Webb pp. 31, 32, 58 top right, 60 left, 60 top right, 61 top left and right, 62 top left, 63 top left, 63 top centre right, 63 top right, 65, 66 top left and right, 66 bottom right.
Wiggins Teape Ltd. p. 94 right.
W.H.O. pp. 60 bottom right, 64, 68 right.
Worthing Museum and Art Gallery p. 41 bottom.
Zoological Society of London p. 101.
After G Maspero, Life in Ancient Egypt and Assyria, London 1892 p. 43 top.

The publishers have made every effort to trace the copyright holders, but if they have inadvertently overlooked any, they will be pleased to make the necessary arrangement at the first opportunity.

Preface

This is a one-year introductory course in biology for the middle years of secondary schools. It is intended ideally as a bridge between an integrated or combined science course and CSE or GCE biology, although it should also provide a valuable foundation of knowledge for those pupils who will take their studies of biology no further.

The term 'lifestyles' refers to the relatively distinctive nutritional patterns found in the living world. The aim of the book is to outline the main features of each lifestyle, to study a variety of organisms that follow each one and to show how organisms with different lifestyles live together in a balanced natural community. In order to provide flexibility in teaching approach and variety in the lessons taught, the book is divided into a large number of self-contained yet interlinked chapters, each representing approximately one week's work.

Inevitably, the approach of this book is a simplification of the diversity of nutritional patterns found in nature, but such simplification makes it possible to touch upon most of the chief features of the biosphere in a short and easily understood course of study. The theme of *Lifestyles* is essentially ecological, providing a coherent base for detailed studies at a later stage within the whole field of biology. In addition, it allows the inclusion of many topics relevant to the everyday life of the individual and of the community. In preparing the text, every opportunity has been taken to refer to the technological applications of biology and their consequences, and thus to equip the pupil to develop a critical understanding of the biological issues that abound in the world today.

Owen Bishop

1 *Viking* looks for life

In July 1976, the lander of *Viking 1* parachuted towards the surface of Mars. Within hours, the first close-up photographs of the Martian surface were being received on Earth, over 300 million kilometres away. The photographs showed a bare, rocky surface. There were no signs of life.

Scientists had not expected to find animals or plants on Mars. They already knew that conditions there are not right for the kinds of living thing we have on Earth. Mars is cold; its surface temperature is nearly always well below freezing point. Its water is almost permanently frozen. Mars has only a thin atmosphere, which allows the deadly ultraviolet radiation from the Sun to reach the Martian surface.

Yet it was believed that some form of life might exist on Mars, in spite of all the difficulties; scientists thought it possible that tiny living things might exist in the soil. The lander had a digging arm to collect soil samples, which were then taken into the lander to be tested in the Biology Pack. This consisted of special equipment for finding out whether the soil contained living organisms. The Biology Pack could do three kinds of experiment. For one experiment, the soil sample was gently warmed and light was shone on it. The atmosphere in the equipment contained carbon dioxide and carbon monoxide, both of which are plentiful in the Martian atmosphere. After five days, the soil was tested to find out whether the gases had been taken into the soil. If this were found to be happening, it would indicate that there were organisms in the soil, using gases from the atmosphere. For the other two experiments, water, with simple food materials dissolved in it, was added to other soil samples. The equipment carried out tests to discover whether the food materials were being changed in any way. Any changes would indicate that there were organisms in the soil.

As so often happens in scientific work, the results of the experiments did not give the clear answers that were hoped for. The experiments did not show *for certain* whether the Martian soil contained living organisms or not. Although the results were hard to explain, they were not useless; much of the information obtained by *Viking 1* and *Viking 2* is helpful for planning future missions to Mars and other planets.

When scientists were designing the Biology Pack, they had no clear idea about the kinds of life they expected to find on Mars. The best they could do was to base their designs on what they already knew about life on Earth.

fig 1.1   Viking 1 *lander nears the end of its eleven-month journey from Earth*

fig 1.2   *The surface of Mars as photographed by* Viking 1. *There are no signs of life.*

During the experiments they planned to give the Martian organisms (if any) the best possible conditions to make them active. Some of these conditions are mentioned in the description of the experiments: warmth, light, atmosphere, water and food materials. On Earth, the factors that are essential for an organism to have a healthy and active life are:

1 Food Food gives *matter* – for making the bodies of organisms, for growth, for reproduction. Food gives *energy* – for all the activities of life. One of the *Viking* experiments gave energy in another form – light; on Earth, plants are able to absorb and use energy in this form (Chapter 2).

2 Water The bodies of organisms contain more than 50 per cent water; most plant matter contains 90 per cent or more. Water is essential to living organisms. Can you think of ways in which they need it?

3 A suitable atmosphere It contains gases that organisms need. Earth's atmosphere contains 21 per cent oxygen; most organisms need oxygen and cannot live if it is absent. Earth's atmosphere contains a small amount (0.03 per cent) of carbon dioxide; green plants need this (Chapter 2). The experimenters thought that Martian organisms *might* need these gases in the atmosphere of Mars, and the Biology Pack was designed to find out whether or not this was true.

4 A suitable temperature On Earth, organisms are most active when the temperature is between about 0°C and about 50°C. They may *survive* temperatures lower or higher than these, but are not usually *active* at temperatures outside this range. For example, many animals hibernate in cold seasons; plants survive as seeds during cold (winter) or heat (in deserts). A medium range of temperatures (20–30°C) is generally the most suitable; this is why it was thought to be a good idea to warm the soil in the *Viking* experiments. Organisms are killed by strong heating; we use high temperature (100°C or more) to kill bacteria (Chapter 12).

5 Freedom from harmful conditions When we put disinfectant on a wound, we make conditions harmful for any bacteria which might try to infect the wound. The disinfectant is a *poison* to the bacteria. Poisons are harmful to life: they create a harmful condition. When detergents pollute streams and rivers, they create conditions that are harmful to water animals. Strong ultraviolet radiation is a harmful condition; on Earth, the atmosphere protects us from most of the Sun's ultraviolet radiation, but the thin atmosphere of Mars gives

fig 1.3 The digging arm of Viking 2 *lander pushes aside a rock before taking a soil sample*

fig 1.4   Mars – a rocky, icy planet with a thin atmosphere containing carbon dioxide, nitrogen and other gases, but no oxygen. What kinds of life, if any, can exist there?

fig 1.5   Earth – a planet with seas and moist soils, with moderate temperatures and an atmosphere containing nitrogen and plentiful oxygen. What kinds of life exist here?

little protection. Small organisms on the soil surface of Mars would be quickly killed. The *Viking* lander pushed aside a piece of rock and dug soil from where the rock had been. It was hoped that the soil would contain live organisms, sheltered from ultraviolet radiation by the rock.

If an organism has food, water, suitable temperature, suitable atmosphere, and there is nothing to harm it, we expect it to remain healthy and to carry out all its normal activities. In the chapters which follow, we shall look at the ways in which the organisms of Earth make use of the good living conditions that exist on most parts of the Earth's surface.

Though there are several millions of different kinds of organism on Earth, there are only about six different ways of living. In the next chapter we shall find out more about one of these six different lifestyles.

2 Leaves in action

Distant photographs of Earth and Mars show no signs of life (p. 3). If we come closer to Mars, we *still* see no signs of life. If we come closer to Earth, we see signs of the activities of organisms. From a few thousand metres above Earth's surface we can see forests which change with the seasons. We can see roads and railways and rivers. There are many signs of activity – especially human activities.

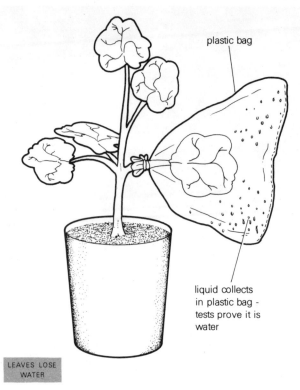

plastic bag

LEAVES LOSE WATER

liquid collects in plastic bag - tests prove it is water

fig 2.2   Evidence 1

fig 2.1   A view of Berkhamsted from the air

When we look at the surface of a leaf, using a lens, we see signs of activity. We see a pattern of veins. They branch and branch again, reaching to all parts of the leaf. Our view of

the veins reminds us of our view of Man's transport systems. Is anything being transported into the leaf? Is anything being transported out of the leaf? Practical tests with living plants can help us answer these questions.

We can demonstrate that plants lose water from their leaves (Evidence 1). Water enters the leaves through the veins, to replace the water that is lost by evaporation. The veins have cells which carry the water – the **xylem** cells (Evidence 2). Examination with a microscope shows that there are small holes in the surface of a leaf. The holes are called **stomata** (one hole is called a **stoma**). The guard cells can change shape to make the hole larger or smaller. When the stomata are wide open, much water vapour is lost from the leaf (Evidence 3). Gases such as oxygen and carbon dioxide can also pass through the stomata, either from the leaf to the atmosphere or from the atmosphere to the leaf.

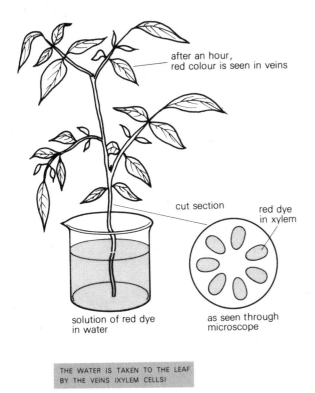

after an hour,
red colour is seen in veins

cut section

red dye
in xylem

solution of red dye
in water

as seen through
microscope

THE WATER IS TAKEN TO THE LEAF
BY THE VEINS (XYLEM CELLS)

fig 2.3   Evidence 2

stomata

guard
cells

fig 2.5   The lower surface of a Tradescantia *leaf, magnified 125 times*

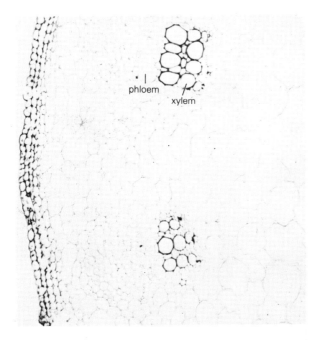

phloem

xylem

fig 2.4   A cross section of the stem of a sunflower plant, highly magnified

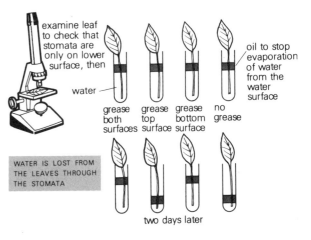

examine leaf
to check that
stomata are
only on lower
surface, then

water

oil to stop
evaporation
of water
from the
water
surface

grease
both
surfaces

grease
top
surface

grease
bottom
surface

no
grease

WATER IS LOST FROM
THE LEAVES THROUGH
THE STOMATA

two days later

fig 2.6   Evidence 3

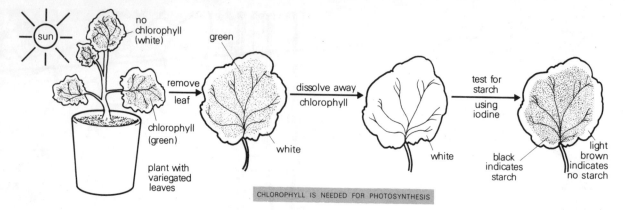

fig 2.7 Evidence 4

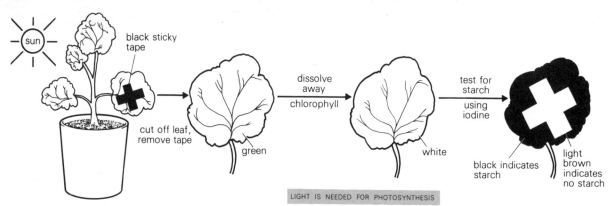

fig 2.8 Evidence 5

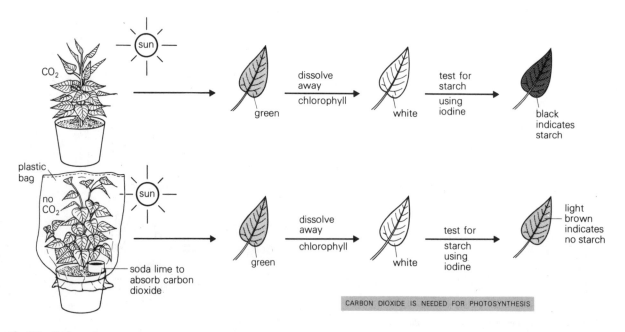

fig 2.9 Evidence 6

6 LIFESTYLES–AN INTRODUCTION TO BIOLOGY

Water from the soil enters the roots of a plant, passes up through the stem (in the xylem cells of its veins) and into the leaves. It carries dissolved mineral salts, which have come from salts in the soil. When the water evaporates from the leaves, these salts are left behind. Some of them are used by the plant, for making substances it needs.

The most important activity of the leaves is **photosynthesis**. This is the making of sugars. In some plants the sugars are then turned into starch. If we test a leaf and find that it contains starch, we can be sure that the leaf has been making sugars. To make sugars, a leaf must contain the green substances called **chlorophyll**. Without this, photosynthesis cannot occur (Evidence 4). Green plants are the only organisms that have chlorophyll, so they are the only organisms that can perform photosynthesis, and make sugar and starch. Photosynthesis is the most distinctive and important feature of the plant's lifestyle.

To be able to photosynthesise, plants need:

1 **Chlorophyll** (Evidence 4)

2 **Light** – a source of *energy* (Evidence 5; see also p. 2, item 1)

3 **Carbon dioxide** – for making sugars (Evidence 6; see also p. 2, item 3) and

4 they also need the other **good conditions**, such as water, a suitable temperature and freedom from harmful conditions (see p. 2, items 2, 4 and 5).

Photosynthesis produces sugars and oxygen (Evidence 7). As they are made, sugars enter the veins and are carried away through the **phloem** cells to other parts of the plant. In bright sunlight, sugars may be made faster than they can be carried away from the leaf. The surplus

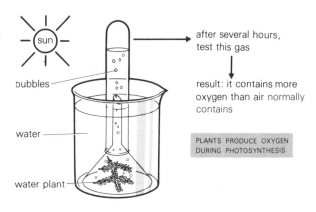

after several hours, test this gas

result: it contains more oxygen than air normally contains

PLANTS PRODUCE OXYGEN DURING PHOTOSYNTHESIS

fig 2.10   Evidence 7

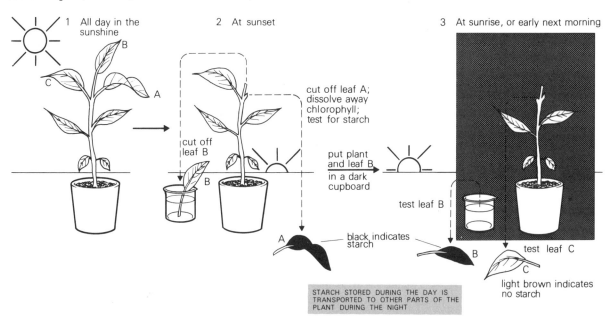

1 All day in the sunshine

2 At sunset

3 At sunrise, or early next morning

cut off leaf A; dissolve away chlorophyll; test for starch

cut off leaf B

put plant and leaf B in a dark cupboard

test leaf B

black indicates starch

STARCH STORED DURING THE DAY IS TRANSPORTED TO OTHER PARTS OF THE PLANT DURING THE NIGHT

test leaf C

light brown indicates no starch

fig 2.11   Evidence 8

sugars are then stored temporarily in the cells of the leaf, dissolved in the water of the cytoplasm. Some kinds of plant convert surplus sugar to starch. This is insoluble, so solid grains of starch accumulate in the cytoplasm. By the end of a long sunny day, there may be a large store of dissolved sugar (and starch grains, if made) in the leaf. During darkness, photosynthesis stops (item 2 above), but the transport of sugars from the leaf continues. Starch itself cannot be carried away, for it is insoluble, so it is re-converted to soluble sugar, which can then enter the vein to be carried away. During darkness, the store of sugar and starch gradually decreases. By morning, little or none is left (Evidence 8).

The plant uses the sugars and mineral salts to make all other substances it needs. These include amino acids, which are then used for making proteins. The leaf is the main part of the plant for making the many complicated chemical substances needed for building the body of the plant. The phloem cells can carry these substances to any part of the plant which needs them, especially:

1   **Growing parts** – tips of roots, tips of shoots, buds, the cambium (Chapter 21)

2   **Storing parts** – tap roots, tubers, bulbs, rhizomes, corms, seeds, fruits (Chapter 20).

In photosynthesis, plants use the energy of sunlight to combine carbon dioxide and water to make sugars (and often to make starch). From sugars and mineral salts they make proteins and all the other substances they need for growing and reproducing. This is the lifestyle of the plants.

3 Cells

Most organisms are made of living units, called cells. Cells are small. We need a microscope to see them clearly. Below is a drawing of a small area of the skin from an onion. The drawing shows 4 whole cells and parts of 11 cells. The cells are drawn about nine times larger than their real size. Cells like these are typical of the cells from which plants are made.

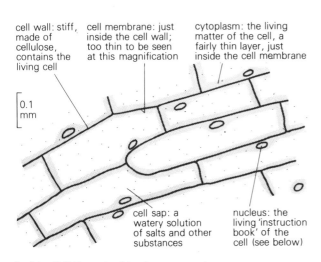

cell wall: stiff, made of cellulose, contains the living cell

cell membrane: just inside the cell wall; too thin to be seen at this magnification

cytoplasm: the living matter of the cell, a fairly thin layer, just inside the cell membrane

0.1 mm

cell sap: a watery solution of salts and other substances

nucleus: the living 'instruction book' of the cell (see below)

fig 3.1 Cells from the skin of an onion scale

The nucleus contains many tiny threads of material. They are called **chromosomes**. The chromosomes carry instructions. These are not words and sentences; these are different kinds of chemical particles, set out in rows on the chromosomes. The rows of particles, like rows of letters on a page, are the instructions to the cell. These instructions say what the cell is to be like as it grows, and what the whole plant is to be like. The instructions decide the colour of its flowers, the height to which it can grow, the flavour of its fruits, and all other **inherited** features of the plant.

Things to do

1 Look at a piece of onion skin, to find cells like those drawn opposite. To make the nucleus and cytoplasm easier to see, add a *drop* of weak iodine solution to the water on the slide. Iodine solution stains them light brown. How many nuclei do you find in each cell?

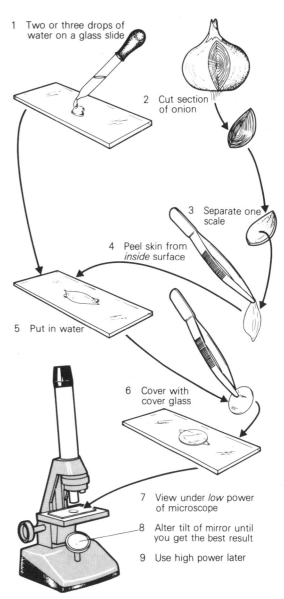

1 Two or three drops of water on a glass slide

2 Cut section of onion

3 Separate one scale

4 Peel skin from *inside* surface

5 Put in water

6 Cover with cover glass

7 View under *low* power of microscope

8 Alter tilt of mirror until you get the best result

9 Use high power later

fig 3.2 How to look at the cells in the skin of an onion scale

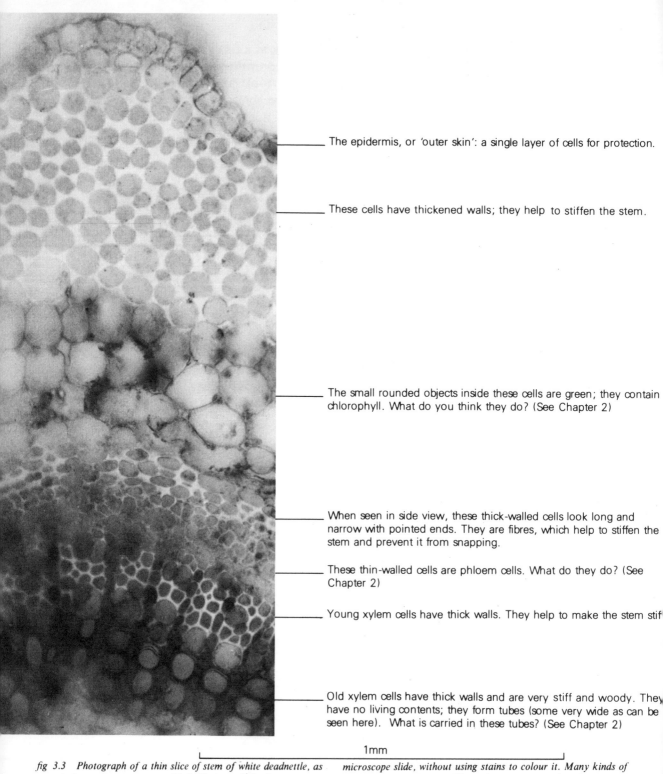

The epidermis, or 'outer skin': a single layer of cells for protection.

These cells have thickened walls; they help to stiffen the stem.

The small rounded objects inside these cells are green; they contain chlorophyll. What do you think they do? (See Chapter 2)

When seen in side view, these thick-walled cells look long and narrow with pointed ends. They are fibres, which help to stiffen the stem and prevent it from snapping.

These thin-walled cells are phloem cells. What do they do? (See Chapter 2)

Young xylem cells have thick walls. They help to make the stem stiff

Old xylem cells have thick walls and are very stiff and woody. They have no living contents; they form tubes (some very wide as can be seen here). What is carried in these tubes? (See Chapter 2)

1mm

fig 3.3 Photograph of a thin slice of stem of white deadnettle, as seen under a microscope, magnified 100 times. This slice was cut by hand, using a razor blade. It was placed in a drop of water on a microscope slide, without using stains to colour it. Many kinds of cell can be seen. You could cut slices just as good as this one.

2   Look at cells from other plants. You can find many different kinds of cells if you look at:

(a) **skin** peeled from leaves (iris) and stalks (rhubarb); or tear a leaf (laurel) and look at the torn edges. Look at flower petals and the thin leaves of mosses.

(b) **pulp** from ripe fruits (tomato, pear, apple, snowberry).

(c) **slices** cut across stems (dead-nettle), twigs (elder), roots (carrot, beetroot), cork and potato. Cut slices across *and along* soft wood (a match-stick).

(d) **hairs** on leaves, stems and fruits (goose-grass, dandelion).

(e) any other plant material that interests you. How many different kinds of cell can you find? In what ways are they different from the 'typical cell' found in the onion skin? Look for cells that do different jobs – supporting (stiffening) cells, transporting cells, protecting cells, photosynthesising cells, food-storing cells. Figure 3.3 shows you what some of these look like.

3   Here are some drawings of cells from plants. Did you find any like these when you worked item 2? If you recognise these cells, say where they are found. If possible, say what their job is. Drawings (b) and (c) are about seventy times larger than life size; the other drawings are about a hundred and fifty times larger than life size.

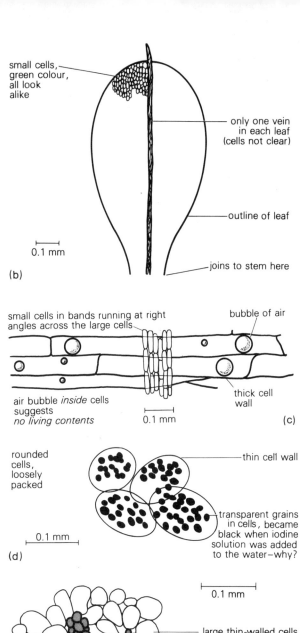

(b)

(c)

(d)

(e)

(f)

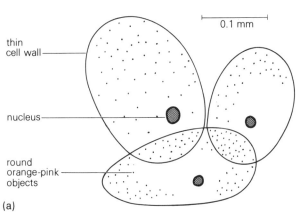

(a)

fig 3.4   (a–f) Can you say where these kinds of cells are found?

4 Simple life

Most plants are built up from a large number of cells. Their cells are of several different kinds. Each kind is specially made to do a different job. If we look at the way in which most animals are built up, we find the same thing, though their cells are different in many ways from the cells we find in plants. For example, a typical cell from an animal has no thick cell wall, and there is no large space inside filled with cell sap.

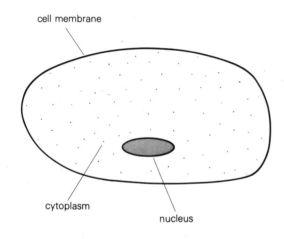

fig 4.1   A typical animal cell. Contrast this with Figure 3.1.

Some plants and animals are made from only one cell. This one cell must perform all the jobs necessary to keep itself alive. The one-celled plants and animals are almost the simplest living things we know, though bacteria and viruses (Chapters 8 and 9) are simpler still.

Let us look at the lifestyles of some of these one-celled organisms. Below, the main features of each are set out in tables, so that you can compare them easily.

Name **PLEUROCOCCUS**	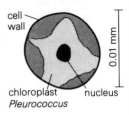
Living place Damp places, especially walls and fences	
*Conditions needed** Light, carbon dioxide	
Source of organic materials † Makes them, by photosynthesis (Chapter 2)	
Way of taking in organic materials None (made inside the cell)	
Way of moving from place to place None	
Special features —	

* *All* organisms also need water, oxygen, suitable temperature and freedom from harmful conditions (Chapter 1).
† Organic materials include sugars, starch, fats, proteins and vitamins.

Name **EUGLENA**	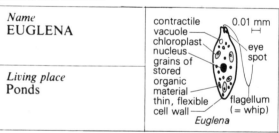
Living place Ponds	
Conditions needed Light, carbon dioxide	
Source of organic materials Makes them, by photosynthesis	
Way of taking in organic materials If kept in darkness, it can absorb soluble organic materials from the water around it	
Way of moving from place to place Lashing action of its flagellum	
Special features Swims towards bright light. Is it a plant? Is it an animal?	

Name **TRICHONYMPHA**	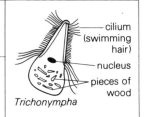 *Trichonympha*
Living place In the gut of termites	

Conditions needed
Organic materials

Source of organic materials
Pieces of wood eaten by termites

Way of taking in organic materials
Projections from rear of cell surround pieces
of wood and take them into the cell. Wood
digested in the cell

Way of moving from place to place
Beating action of its cilia

Special features
Very few organisms can digest wood

Name **SACCHARO- MYCES (YEAST)**	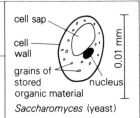 *Saccharomyces* (yeast)
Living place Skin of very ripe fruit, rotting or- ganic material	

Conditions needed
Soluble organic materials

Source of organic materials
Sugary materials absorbed; can make others
from these

Way of taking in organic materials
Absorbs them from solution, through its
surface

Way of moving from place to place
None

Special features
Very useful to Man for making bread, wine,
beer

Name **PARAMECIUM**	*Paramecium*
Living place Ponds	

Conditions needed
Organic materials

Way of obtaining organic materials
Food (mainly bacteria) caught in mouth
groove; then into food vacuoles, where it is
digested

Way of moving from place to place
Beating action of its cilia

Special features
Avoids harmful conditions – high and low
temperature, harmful substances in water,
enemies

Name **DIDINIUM**	*Didinium*
Living place Ponds	

Conditions needed
Organic materials

Source of organic materials
Paramecium

Way of taking in organic materials
Catches with tentacle, swallows whole, digests
inside cell

Way of moving from place to place
Beating action of cilia

Special features

Name **CLOSTERIUM**	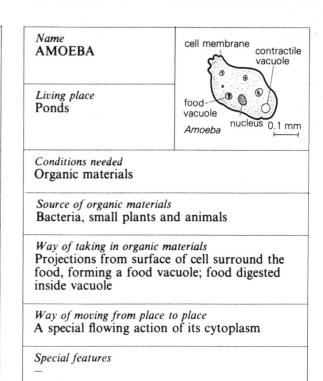
Living place Ponds	

Conditions needed
Light, carbon dioxide

Source of organic materials
Makes them, by photosynthesis

Way of taking in organic materials
None

Way of moving from place to place
Can glide along slowly

Special features
An example of a large group, the *desmids*

Name **AMOEBA**	
Living place Ponds	

Conditions needed
Organic materials

Source of organic materials
Bacteria, small plants and animals

Way of taking in organic materials
Projections from surface of cell surround the food, forming a food vacuole; food digested inside vacuole

Way of moving from place to place
A special flowing action of its cytoplasm

Special features
—

Name **SCENEDESMUS**	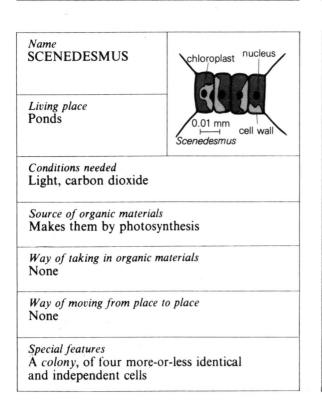
Living place Ponds	

Conditions needed
Light, carbon dioxide

Source of organic materials
Makes them by photosynthesis

Way of taking in organic materials
None

Way of moving from place to place
None

Special features
A *colony*, of four more-or-less identical and independent cells

---

Name **PODOPHYRA**	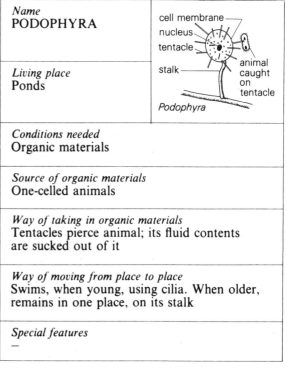
Living place Ponds	

Conditions needed
Organic materials

Source of organic materials
One-celled animals

Way of taking in organic materials
Tentacles pierce animal; its fluid contents are sucked out of it

Way of moving from place to place
Swims, when young, using cilia. When older, remains in one place, on its stalk

Special features
—

Things to do

1   Examine living specimens of some of the above organisms, using a microscope.

2   Draw out some blank tables like those opposite. Then fill them in with details of some of the one-celled organisms from this list: *Arcella, Chlamydomonas, Chlorella, Colpidium,* diatoms, *Difflugia, Globigerina, Noctiluca, Peranema, Spirogyra, Spirostomum, Stentor, Stylonichia, Volvox, Vorticella.* You may be able to examine slides or cultures of some of these, too.

3   Find out the job of the contractile vacuole, found in *some* of these organisms.

4   There are many variations in the lifestyles of the one-celled organisms, but we can sort them out under five main headings:

(a)   **Plants** – have chlorophyll, make their own organic materials by photosynthesis (Chapter 2). These are the **primary producers** of organic materials.

(b)   **Herbivores** – animals that feed on plants; this is how they get the organic materials they need.

(c)   **Carnivores** – animals that feed on other animals; this is how they get the organic materials they need; the animals they feed on probably got them from plants.

(d)   **Omnivores** – animals that feed on plants *and* other animals.

(e)   **Decomposers** – organisms such as bacteria and fungi that obtain organic material from animals and plants, usually after they are dead, often in some soluble form.

Write out the five headings and list each of the one-celled animals you have studied under one of these headings (are there any that are hard to place?). You have now classified these organisms according to their lifestyle.

5   Organisms can also be classified into families or other groups of closely-related organisms. Which of the organisms you have studied can be classified as (a) algae, (b) protozoa, (c) fungi?

5   Seeds for survival

In some parts of the world, plants live in an ideal climate:

In other parts of the world, there are times when survival is difficult:

fig 5.2   A garden in Khartoum, Sudan, Africa. Only weekly irrigation with water pumped from the river Nile keeps these plants alive. Outside the city area is semi-desert. Annual rainfall is only 125 mm; this falls during three months and is followed by nine months of dry season, when no rain falls.

fig 5.3   Snowdrops in England in early spring. During winter, sunlight is weak, days are short and temperatures are low. Water may be frozen and not available to plants. Leaves may be killed by frost. These plants survived the winter as corms, sheltered beneath the soil (Chapters 23 and 24).

fig 5.1   Flowers of Morning Glory in a garden in the Highlands of New Guinea. The sun shines daily, and almost every evening there is a shower of rain to keep the soil moist. The temperature is never too hot and never too cold.

If the climate includes long periods of drought, or great heat, or severe cold, seeds can help the plant survive. Inside a seed, the young plant (or embryo) is inactive, but it can remain alive for a long time. It needs no supply of water. It can survive high temperatures and low temperatures much more easily than an actively growing or fully-grown plant can. The embryo is able to wait until conditions are suitable for it to begin to grow into a plant.

In the northern parts of the world (including Britain), summer is the best season for growing plants. In desert areas the growing season is the short period during which the rains fall.

Annual plants live for only one growing season. Young annual plants grow from seed at the beginning of the season and quickly reach full size. They form flowers and then fruits. The fruits contain seeds. Ripe fruits or seeds are scattered from the plant towards the end of the growing season. Then the annual plant dies completely. Only the seeds are left to survive the winter or dry season, until the next growing season begins.

When the seeds of annual plants are newly made, conditions are still suitable for growth. If the seeds began to germinate then, towards the end of the season, they might grow well at first but then be killed by frost or drought. They might not survive. There are ways of avoiding this problem. We often find that seeds will *not* germinate when they are newly made – even if they are put in ideal growing conditions. We say that the seeds are **dormant**, a word that means 'sleeping'. They remain dormant for many weeks or months. Usually their dormancy is over by the beginning of the next growing season. Then they can germinate safely, and survive.

With some kinds of plant we find that seeds remain dormant for periods of different lengths, from a few months up to several years. After a plant has produced a batch of seeds, a few of these germinate each year. If there is an especially bad year, with long drought or intensely cold spells, and most of the seedlings of

fig 5.4   The ash tree produces its fruits in clusters, like bunches of keys. Each ash 'key' has a wing to delay its fall to the ground. It can be blown several metres away from the tree, to an area in which it finds the right conditions for germinating. But, even if conditions are right, it remains dormant for two years or more.

that year die, some dormant seeds still remain safely in the soil, to germinate in following years.

Dormancy is caused in various ways in different kinds of seed. The seeds of some desert plants remain dormant because a substance in the seed or fruit prevents the embryo from growing. When the rains come, *and if there is*

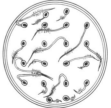 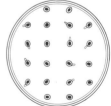

A   18 germinated,
    2 not germinated

B   10 germinated (just),
    10 not germinated

fig 5.5   Two batches of tomato seeds were placed in dishes to germinate. Those in dish A were supplied with plain water; those in dish B were supplied with water to which twenty-five per cent juice from fresh tomatoes had been added. Four days later they looked like this. Can you explain the difference in the way the two batches have germinated? What is the importance of this effect for tomatoes growing wild?

enough rain, the substance is washed away. Then the embryo can begin to grow. A light shower alone is not enough to wash all of the substance away. This is why the seeds do not germinate until the amount of rain is sufficient to make the soil really wet for the newly growing seedlings.

The seeds of some plants of northern areas remain dormant until after they have been exposed to low temperature for a long time. This is why they can not germinate in autumn. During winter they are cold for several months. After this they are not dormant. Then, when the temperature rises enough in spring or early summer, they begin to grow.

Seeds are an essential part of the life story of annual plants, and an important part of the life story of many other kinds of plant. Seeds help plants to survive unfavourable times of year. Seeds are important in other ways too. The embryo is a young plant in compact form. It can be easily transported from the parent plant. It can be carried to other areas where it *may* find conditions that are suitable for its growth. Conditions there *may* be better than those in the area in which its parent grew. By means of seeds, plants are able to spread to all suitable areas, so obtaining the maximum possible living space.

fig 5.7   *The feathery tufts on the fruits of dandelion catch the slightest wind and they are carried kilometres away from the parent plant*

fig 5.6   *Ripe apples in autumn (top). The seeds (bottom) are dormant. They will not germinate in autumn. They need about three months of cold weather before they leave the dormant state.*

fig 5.8   *The bright red colour of hawthorn fruits makes them an attractive food for birds, especially if winter is cold and long. The seeds are eaten, for they are inside the fruit, but they pass unharmed through the digestive system of the bird. When they are dropped, they may be many kilometres from the parent bush.*

Things to do

1   In autumn, collect seeds and try to germinate them immediately. Sow the remainder in small pots of moist soil: leave them for two weeks to absorb water, then transfer them to a cold place (refrigerator or deep-freeze) for periods ranging from two weeks to four months.
After the cold treatment, transfer the pots to a warm place (laboratory) to find out whether the seeds germinate, or whether they are still dormant. For each type of seed, find the least period of cold needed to end dormancy. Seeds worth trying include: sycamore, apple, rose, hawthorn, ash, hazel.

2   In what ways do *animals* try to avoid the worst effects of (a) a cold season, or (b) a dry season?

3   The plants in Figure 5.1 are in an ideal climate *but* they have at least one harmful condition acting against them. Look again at the photograph to discover what it is.

6 The fungi

Yeast is a fungus (Chapter 4), but it is not a typical one. Yeast consists of a single oval cell, but almost all other fungi consist of a mass of thin, branching threads, which spread through the material on which the fungus is living.

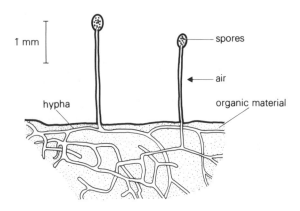

fig 6.1 Mucor, or pin-mould, grows on jam and bread. The hyphae are tubes of a cellulose-like material. The tube is lined with cytoplasm, with nuclei scattered in it. The centre of the tube contains cell sap. Mucor reproduces by making spores (see below).

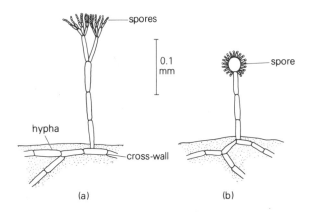

fig 6.2 (a) Penicillium (b) Aspergillus. These blue moulds live on a wide variety of organic materials – bread, jam, cheese, leather, dead leaves. Unlike Mucor, their hyphae consist of cells joined end-to-end, with cross-walls between.

All fungi obtain the organic materials they need by taking them from their surroundings. They have no chlorophyll so they are not able to photosynthesise and make the organic materials for themselves. Their hyphae spread through their food material. Soluble organic substances, such as sugars, are absorbed into the hyphae for use by the fungus. Water and mineral salts are taken in, too. Insoluble organic substances, such as starch, cellulose and proteins, are first made soluble by the action of digestive agents (enzymes) which are given out by the hyphae. These substances digest the organic materials in the region around the fungus. When they have been converted to soluble substances, they are absorbed by the hyphae. Given suitable temperature, oxygen in the air, enough water (a reasonably damp place) and absence of harmful conditions, fungi can live on a supply of suitable organic matter. Here are some places where fungi live well:

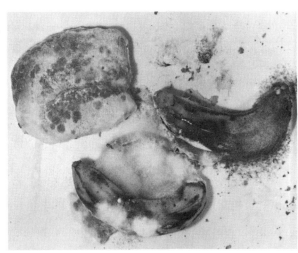

fig 6.3 Bread, fruit and other foodstuffs

fig 6.4 Piles of dead leaves, including compost heaps, if they are in a damp, shady place

fig 6.5 Soil, if rich with dead plant and animal matter. The picture shows a 'fairy ring'.

By their action of digesting and absorbing organic materials, fungi act as **decomposers**. As we shall see later, they are an essential part of the natural cycle of growth and decay.

Fungi reproduce themselves by forming **spores**. These are usually small and light. They are carried long distances by wind, but some are carried away on the bodies of passing insects. When the spores reach a place where the living conditions are correct, they germinate. New hyphae grow from them and spread into the surrounding food materials. A fungus such as *Mucor* makes its spores in a round body called a **sporangium**. Other fungi, such as

fig 6.6 Dead wood. A fungus is the cause of dry rot. Dry rot damages the structural timber of houses. It needs dampness (for example, a leaking roof) to begin to grow, but the infected timber becomes dry and cracked and crumbles away to powder when touched. Dry rot is hard to eradicate; the remedy is to cut away and burn all the affected timber.

Penicillium and *Aspergillus*, form spores in chains at the ends of special hyphae. The spore-forming bodies we most often notice are the large mushrooms and toadstools made by members of one group of fungi. If the autumn is warm and damp it is easy to find toadstools, bracket-fungi, puff-balls and other large fruiting bodies in a wide variety of shapes, sizes and colours. These produce millions of spores each, though few of these spores ever arrive at a place where they can germinate. Why do most spores never get to a favourable place? What happens to them?

Things to do

Here are some ways of obtaining fungi, so that you can find out more about them:

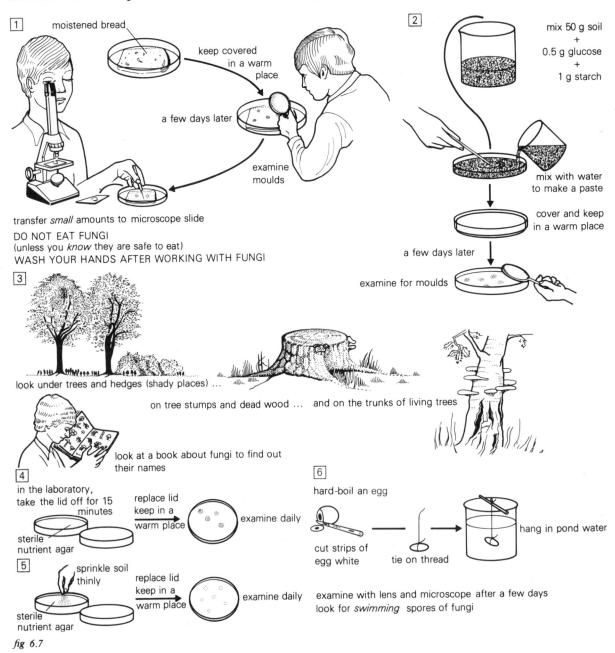

1. moistened bread
keep covered in a warm place
a few days later
examine moulds
transfer *small* amounts to microscope slide
DO NOT EAT FUNGI
(unless you *know* they are safe to eat)
WASH YOUR HANDS AFTER WORKING WITH FUNGI

2. mix 50 g soil + 0.5 g glucose + 1 g starch
mix with water to make a paste
cover and keep in a warm place
a few days later
examine for moulds

3. look under trees and hedges (shady places) ...
on tree stumps and dead wood ...
and on the trunks of living trees
look at a book about fungi to find out their names

4. in the laboratory, take the lid off for 15 minutes
replace lid keep in a warm place
examine daily
sterile nutrient agar

5. sprinkle soil thinly
replace lid keep in a warm place
examine daily
sterile nutrient agar

6. hard-boil an egg
cut strips of egg white
tie on thread
hang in pond water
examine with lens and microscope after a few days
look for *swimming* spores of fungi

fig 6.7

Robert Koch

Robert KOCH, a German bacteriologist, lived from 1843 to 1910. He invented many techniques useful for examining and studying bacteria, and has been called 'The Father of Bacteriology'. He invented methods of preparing 'smears' of bacteria on microscope slides. He found ways of staining them to make them easily visible, so that they could be identified more easily. He thought of the idea of growing bacteria in dishes, on a layer of sterile gelatine jelly, with added food materials. This technique makes it much easier to study bacteria, and is still used today, though agar (a jelly-like material from seaweeds) is now used instead of gelatine. Using his techniques, he was the first person to discover the bacteria causing cholera, tuberculosis, typhoid and several other diseases.

7 Using the decomposers

The decomposers include fungi and bacteria. In nature they perform an essential task by decomposing materials derived from plants and animals. The elements from which these materials are made are then returned to the soil and may be used again (Chapter 32). Unfortunately, decomposers also spoil and destroy our food, our clothing and our buildings if we do not take proper precautions to prevent this. But we have learned how to make good use of *some* of their decomposing activities.

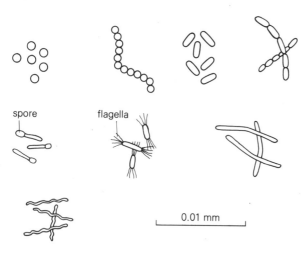

fig 7.1 *Bacteria have many different shapes. As the scale shows, they are all very small. Even under a microscope it is difficult to see more detail than is given in these drawings.*

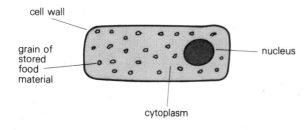

fig 7.2 *A typical bacteria cell has a simple structure. Some bacteria have a layer of slime outside the cell wall.*

When a yeast cell absorbs sugars it uses them partly to provide itself with energy for living. To do this, it also absorbs oxygen. Then the sugars are broken down (or decomposed) inside the yeast cell:

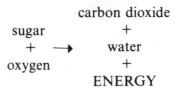

$$\text{sugar} + \text{oxygen} \longrightarrow \text{carbon dioxide} + \text{water} + \text{ENERGY}$$

This process is called **respiration**. It occurs in the cells of almost all organisms (including human cells), as a way of releasing the chemical energy stored in food materials. We make use of yeast's respiration when we prepare bread. First we make dough, by mixing flour, water and other ingredients. Flour consists mainly of starch, but enzymes present in the flour convert some of the starch to sugar when the flour is moistened. Also, we often add sugar or some other sugar-containing substance (for example, milk or honey) to the dough to give the yeast a good supply of soluble food. Then we add yeast to the dough and knead it vigorously to mix them well and to make the dough springy and elastic. The dough is then left in a warm place for an hour or two. The yeast respires. It gives off carbon dioxide, and millions of tiny bubbles of this gas are formed throughout the dough. The extra volume of gas bubbles makes the dough 'rise'; its volume increases to twice its original size or more. When the dough is put in the oven, the yeast is killed by the heat, but the bubbles of gas expand in size even further, giving the bread a pleasantly light and porous texture. This texture is made permanent because the dough becomes stiffer and less springy when it is cooked.

Yeast can respire in another way when no oxygen is available:

$$\text{sugar} \longrightarrow \text{carbon dioxide} + \text{ethanol} + \text{ENERGY}$$

fig 7.3 Making wine at home. The jar on the left contains a mixture of boiled rice, sugar and other ingredients in water. The jar on the right contains a mixture of plum juice, orange juice, sugar, other ingredients and water. Yeast has been added to each jar and fermentation is proceeding. Note the glass traps at the top of each jar; these let out the carbon dioxide as it bubbles from the fermenting juices. The traps prevent air from entering the jars so fermentation is anaerobic. They also prevent small fruit-flies from reaching the wine – these might carry vinegar bacteria to the juices which would then be fermented to vinegar.

This is **anaerobic** (= 'no air') respiration, the basis of **fermentation.** When ripe fruits (for example, grapes) are crushed, their juices, which contain sugars, can be squeezed out. Wild yeasts are living on the skin of the fruits. If the juice is left in a covered container (to exclude air), the yeasts ferment the sugars, producing ethanol. This is the basis of making wine. To make beer, we germinate barley seeds. This converts their stored starch to sugars. The germinated seed is roasted to kill it and to add flavour – this gives us malt. The malt is washed with water to extract the sugar and flavour, giving wort. Flavourings such as hop extract are added to this. Then yeast is added to the wort, to ferment the sugar to ethanol. If we ferment extracts from cheap or surplus crop plants, we can extract pure ethanol by distilling the fermented mixture. This is one way in which ethanol is made industrially, for use as a solvent and in the chemical industry. Ethanol can be used as fuel for internal combustion engines; when supplies of natural oil run out, it could become an important substitute for petrol.

We use other organisms for different types of fermentation. When the vinegar bacterium ferments sugars, it produces acetic acid instead of ethanol. It is used to ferment grape juice (giving wine vinegar) or barley wort (giving malt vinegar). The acetic acid made by the bacterium gives vinegar its pleasantly sharp flavour. *Lactobacillus*, a common bacterium in milk, ferments sugar in milk to lactic acid. This gives the milk a sour taste, unpleasant with cornflakes or in a cup of tea, but pleasant if we make the milk into cottage cheese. The lactic acid prevents other organisms from living in the milk and decomposing it – it stops them from perhaps producing unpleasant flavours or even poisonous substances. In this way we can keep the milk (as cheese) for a few days longer, and it is still safe to eat.

For centuries we have used decomposers for preserving food: by making milk into cheese or yoghurt, by making cabbage into sauerkraut, by making fruit juices into wine. Food can be preserved when it is plentiful for use when it is scarce. We have studied the decomposing organisms and found out the best and safest ways of using them. *Finding out* about these things is *science*: when we *use* our knowledge for our benefit, that is *technology*. The use of fermentation is one of today's major technologies. We are the only animals able to study the world around us *and then make good use of what we discover*. Ours is a special kind of lifestyle – that of the **technological animal**.

fig 7.4 *Early fermentation technology*

Things to do

1 We use the decomposers in many ways. Find out more about the processes listed below and, if possible, set up demonstrations of these processes for yourself: making yoghurt, making cheese, making butter, making vitamin B tablets, composting garden refuse, production of biogas, retting of flax, production of silage, pickling of foods such as cabbage (sauerkraut), walnuts, plums or onions.

2 Try to grow some of the decomposers that occur in milk, yoghurt, cheese and garden compost, by making cultures of them on dishes of nutrient agar.

Louis Pasteur

Louis PASTEUR, a French scientist, lived from 1822 to 1895. His first scientific work was to study the structure of crystals. This led him to many investigations of fermentation. In those days most people believed that when matter decayed or fermented the microscopic organisms causing the decay were *created* in the matter. Pasteur showed that this is not true – living organisms can come only *from living organisms*. He proved that organisms or their spores are carried in the air and, when they land on suitable matter, can multiply there and cause it to decay. He showed that the souring of milk is caused by bacteria, and that bacteria are involved in the decomposition of silage. He also showed that these bacteria could be killed by high temperatures. Brewers and winemakers asked Pasteur to help them find out why their fermentations sometimes went wrong. He discovered the organisms which were causing their troubles, and suggested how they could be eliminated. He invented pasteurisation – a way of heating wine to kill the organisms in it that were giving it a sour taste. Today, this treatment is used for milk, to help keep it in a fresh, safe condition for longer than it will keep as raw milk. Pasteur was asked to study the diseases of silkworms, since silk production was then an important industry in France. He was successful in finding the causes of several diseases of silkworms and discovering ways of preventing them. His work helped make people believe that most diseases are caused by 'germs'. He also invented methods of vaccination against two serious diseases: *anthrax*, a fatal disease of cattle, sheep and horses, and *rabies*, a horrible and fatal disease that can attack humans.

8 Bacteria in the soil

If we put soil in a jar containing water, stir it, and then leave it, the soil settles in layers (Figure 8.1) The largest gravel particles (including stones and pebbles, if any) fall immediately to the bottom of the jar. Coarse particles of sand settle along with these. A second or so later, the finer sand particles settle on top of the coarse sand. The next to settle are the particles of silt. Clay particles are the finest of all. They settle very slowly. They may take several hours or even days to settle out completely. Soil also contains small animals, which may float out when the soil is put in water. It also contains **raw humus**, which consists of dead animals or parts of animals and parts of plants, such as pieces of root, seeds, and pieces of leaf and stem. Some pieces of raw humus rest on top of the clay and others remain floating in the water.

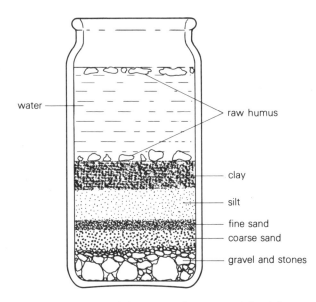

fig 8.1 *When soil is shaken or stirred in water and then left to stand, it settles into layers, like these. Try different soils from your local area to see whether they look like this.*

The soil bacteria are far too small to be seen without using a powerful microscope. Even though there are millions of bacteria in a teaspoonful of garden soil it is not easy to see them (even *with* a microscope) among the tiny particles of clay and the smallest pieces of raw humus. The best way to prove that soil contains bacteria, is to grow them on nutrient jelly (Figure 6.7, part 5).

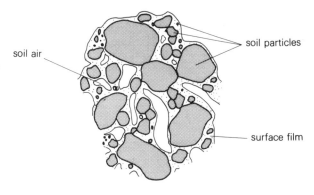

fig 8.2 *Here we see soil highly magnified. Particles of different sizes and kinds are mixed together. They are covered with a thin layer of humus (not shown) and a surface film of water. Between the particles are spaces containing air.*

Soil bacteria and other decomposers (mainly fungi and one-celled animals) live in the surface film of moisture that surrounds each soil particle (Figure 8.2). They obtain air from the spaces between the larger particles. They feed on raw humus. Often they digest the humus by sending out enzymes and absorbing the digested materials. As they break down the raw humus, many soluble compounds are produced. The raw humus is gradually converted to a black sticky substance, called **humus**. Each soil particle becomes covered with a thin layer of humus, which gives soil its blackish or dark-brown colour. The slight stickiness of humus helps join the soil particles together, giving the soil a good texture. The humus also helps the soil to remain reasonably damp in dry weather. Since the humus has been formed by the breakdown of raw humus, it is rich in mineral salts. These dissolve in the water of the surface film and can then be taken into the roots of plants. In this way, essential mineral elements, such as potassium, sulphur and magnesium, are returned to the soil to be used once again by plants.

Nitrogen is an important mineral element, for it is found in all proteins and many of the other organic substances of which living matter is made. Figure 8.3 shows that most of the nitrogen in humus is in the form of ammonium

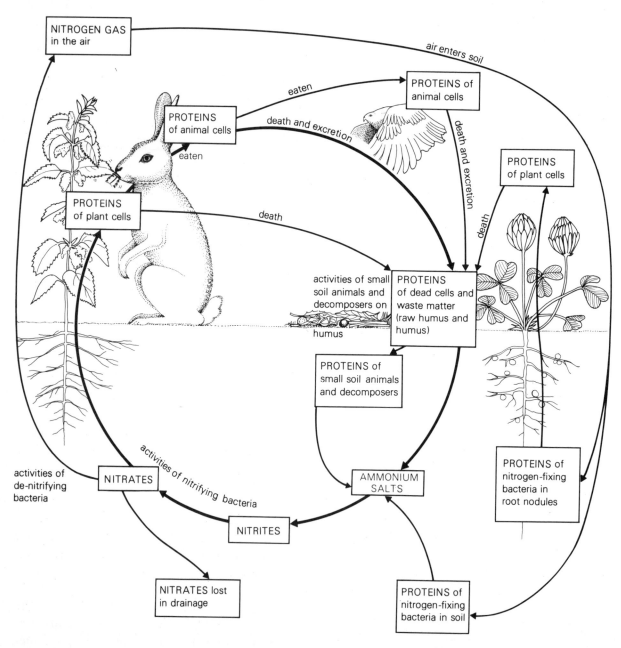

fig 8.3 The nitrogen cycle in nature. Capital letters show the various nitrogen-containing compounds that occur in the cycle. Follow around the main nitrogen cycle (thick arrows) to see what changes occur between the time when nitrates are absorbed from the soil by plants and the time when they are made available in the soil by nitrifying bacteria, ready to be absorbed again. Then follow the thin arrows to see how nitrogen is lost from the soil and how the soil can gain nitrogen.

compounds. Ammonium compounds are not easily taken in by plants, but they can be used by some kinds of bacteria – the **nitrifying bacteria**. Some of these convert the ammonium compounds to nitrites; others convert the nitrites to nitrates. Nitrates are very soluble and are easily absorbed by plants. This completes the **nitrogen cycle**.

If soil is flooded or becomes waterlogged for long periods, all the spaces between the soil particles become filled with water. There is no air in the soil. Such conditions favour the **denitrifying bacteria**. These do not need oxygen for their respiration, so they can multiply at the expense of other kinds of bacteria. Denitrifying bacteria convert nitrate to nitrogen, which escapes from the soil into the atmosphere. This is why flooding and poor drainage decrease soil fertility.

Although almost 80 per cent of the atmosphere is nitrogen gas, this abundant source of nitrogen cannot be used by plants. But certain types of bacteria *can* use nitrogen gas – these are the **nitrogen fixing bacteria**. Some of these live in the surface film of the soil. They absorb dissolved nitrogen and use it to make the proteins they need, as well as other substances. Eventually these nitrogen-containing compounds enter the nitrogen cycle (see Figure 8.3). By this route, nitrogen from the atmosphere becomes available for use by plants. Other kinds of nitrogen-fixing bacteria live inside small swellings (nodules, Figure 8.4) on

fig 8.5 (above)   Moorland soils often lack nitrogen because of heavy rainfall that washes away soluble compounds, and because of the action of denitrifying bacteria in the marshy areas. Gorse is a member of the legume family; the nitrogen-fixing bacteria in its root nodules help it to live in soils that have a very low nitrogen content.

fig 8.6 (below)   Sundew lives in marshy areas where the soil contains little nitrogen. Sundew plants can trap small insects on sticky hairs on their leaves. Juices from the leaves digest the soft parts of the insects. The proteins are digested to give soluble nitrogen-containing compounds. The leaf absorbs nitrogen-containing compounds and in this way the plant gains nitrogen.

fig 8.4   Like most of the members of the legume family, this runner bean plant has nodules on its roots. The nodules each contain millions of nitrogen-fixing bacteria.

root nodules

the roots of certain kinds of plant. These plants gain some of the nitrogen-containing compounds for their own use. When the plants die, the nitrogen-containing compounds enter the nitrogen cycle, to become available to all kinds of plant.

So, we can see how bacteria are essential in keeping soil fertile. They help recycle mineral elements so that they may be used over and over again.

fig 8.9 (above) Making a compost heap in the garden: the heap is built up from layers of waste plant matter. Often a thin layer of powdered chemicals, rich in nitrogen, is sprinkled between the layers. This provides an extra supply of nitrogen-containing compounds and also stimulates the bacteria and other decomposers to break down part of the plant matter to form humus. When the compost is sufficiently decomposed, it is dug into the garden beds. What are the advantages of making and using a compost heap?

fig 8.7 Some of the meadows here are sown with a mixture of grasses and legume plants, such as clover. These provide hay for feeding farm animals. After a few years, the field is ploughed and used for growing other farm crops. When the remains of the clover plants are ploughed into the soil, the nitrogen they have gained from their nodule bacteria is added to the soil. The fertility of the soil is increased.

fig 8.8 (below) Nowadays, farmers try to keep their soil fertile by adding artificial chemical fertilisers. Unlike manure, these do not provide the soil with humus. The soil often becomes dusty and easily blows away – especially if the hedges and trees have been removed to make the fields larger. Why do modern farmers wish to make their fields larger than they used to be in earlier centuries? What are the dangers of cutting down hedges and of making great use of artificial fertilisers?

9 Bacteria and disease

The bacteria that live in soil or in ponds and rivers do a useful job. They are decomposers (Chapter 8). Several other kinds of bacteria have a lifestyle that is quite different from this. For example, think of the bacteria that live in or on your body:

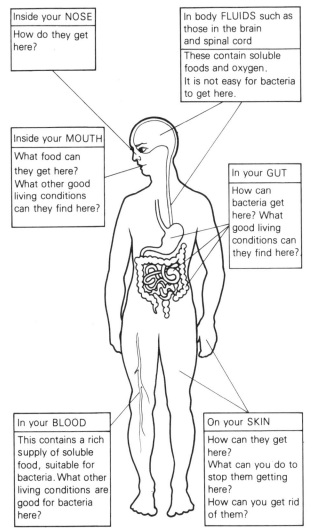

Inside your NOSE
How do they get here?

In body FLUIDS such as those in the brain and spinal cord
These contain soluble foods and oxygen. It is not easy for bacteria to get here.

Inside your MOUTH
What food can they get here? What other good living conditions can they find here?

In your GUT
How can bacteria get here? What good living conditions can they find here?

In your BLOOD
This contains a rich supply of soluble food, suitable for bacteria. What other living conditions are good for bacteria here?

On your SKIN
How can they get here? What can you do to stop them getting here? How can you get rid of them?

fig 9.1

Most of the bacteria of your mouth, gut and skin are harmless. Some are even useful. In your mouth, for example, the harmless bacteria that normally live there make it difficult for harmful bacteria from outside to settle in your mouth and multiply there. The harmless bacteria outnumber the harmful ones. In the competition for food and living-space, the harmful bacteria are the losers. The natural population of harmless bacteria in the mouth protects us from the invaders. We find many kinds of bacteria in the colon (a part of the gut). These make useful amounts of vitamins of the B group. The vitamins are absorbed into our bodies. We and these bacteria live in partnership: each partner helps the other. What do these bacteria get in return for the vitamins they give us?

A few types of bacteria are able to enter the body and to multiply there — and most of these cause harm. They damage the cells in the parts of the body in which they are living. They may make poisonous substances, which may spread throughout the body. When these bacteria are inside us in large numbers, we do not feel well. We may have fever or a sore throat; we may develop spots on the skin or a skin rash; we may feel dizzy or we may vomit: we may even become paralysed or die. The effects we feel depend on what kind of bacterium is living inside us. One or more of these effects can stop us from feeling comfortable or at ease. In short, we feel *dis*-ease because of the effects of the disease-causing bacteria inside us. Bacteria are not the only kinds of organism to cause disease:

fig 9.2 (a) Some fungi are able to invade the skin, causing skin diseases such as Athlete's Foot (shown here) and ringworm.

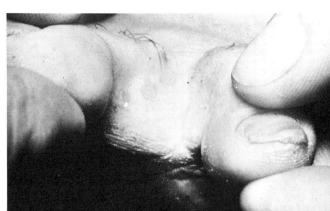

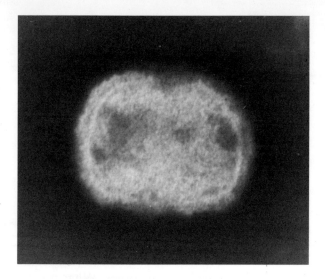

fig 9.2 (b) Viruses are the smallest and simplest known organisms. They all live as parasites, causing diseases such as colds, influenza, poliomyelitis, measles, smallpox and mumps.

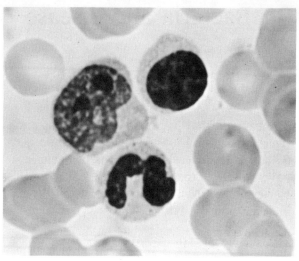

fig 9.3 Human blood magnified 2500 times. It consists of a liquid (plasma) in which several different kinds of cell are suspended. Use the drawings and descriptions in. Figure 9.4 to help you identify the different kinds of blood cell shown in this photograph.

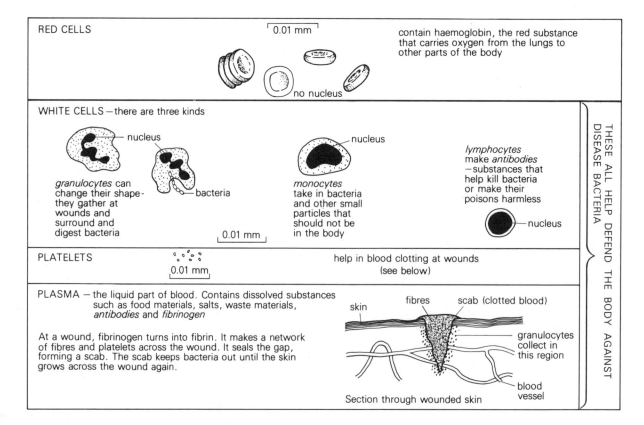

| RED CELLS | 0.01 mm | contain haemoglobin, the red substance that carries oxygen from the lungs to other parts of the body |

no nucleus

WHITE CELLS—there are three kinds

nucleus

granulocytes can change their shape— they gather at wounds and surround and digest bacteria

bacteria

0.01 mm

nucleus

monocytes take in bacteria and other small particles that should not be in the body

lymphocytes make antibodies —substances that help kill bacteria or make their poisons harmless

nucleus

PLATELETS   0.01 mm   help in blood clotting at wounds (see below)

PLASMA — the liquid part of blood. Contains dissolved substances such as food materials, salts, waste materials, antibodies and fibrinogen

At a wound, fibrinogen turns into fibrin. It makes a network of fibres and platelets across the wound. It seals the gap, forming a scab. The scab keeps bacteria out until the skin grows across the wound again.

skin   fibres   scab (clotted blood)

granulocytes collect in this region

blood vessel

Section through wounded skin

THESE ALL HELP DEFEND THE BODY AGAINST DISEASE BACTERIA

fig 9.4 Blood consists of red cells, white cells, platelets and plasma. This diagram shows what each of these looks like and what it does. All except the red cells help defend the body against attack by harmful bacteria.

Bacteria, fungi, viruses and small animals that live inside the body of an organism and take their food from its tissues or fluids are called **parasites**. We shall find out more about the parasite lifestyle in Chapter 18.

Things to do

1   Collect information about disease bacteria. Make lists of:
(a) places where disease bacteria might be found
(b) ways in which they can enter the body
(c) places inside the body in which they can multiply and cause disease
(d) the names of some of the diseases they cause.

2   Find out which are the commonest diseases and whether each is caused by a bacterium, a virus or by some other kind of parasitic organism.

3   Humans are not the only organisms to suffer disease. Find out about diseases of other animals (especially pet animals and farm animals) and of plants.

Joseph Lister

Joseph LISTER was born in England in 1827. He became Professor of Surgery in Glasgow, Edinburgh and London. At that time, a surgical operation was very dangerous for the patient. There was a high risk of infection of the surgical wounds and a high percentage of patients died as a result of being operated upon. Lister studied the causes of this and came to the conclusion that bacteria were being allowed to reach the wounds. He began the practice of **antiseptic surgery**. He said that the operating theatre must be kept clean; that surgeons and their assistants must wear clean clothes when operating; that they must wash their hands before operating and that all surgical instruments must be carefully cleaned before use. He introduced the use of carbolic acid as an antiseptic. Solutions of carbolic acid were used to disinfect the surgeon's hands and the surgical instruments. Operation wounds were washed with carbolic acid to prevent infection. Lister also invented a special spray which was used to spray carbolic acid droplets into the air around the operating table. As a result of these new ideas (except for the spray, which was found to be unnecessary when all the other precautions were taken) the number of deaths resulting from operations decreased considerably. Lord Joseph Lister received his peerage in 1897 and died in 1912.

10 Our water supply

The average human body needs two litres of water each day. We do not need to *drink* the whole of this. The average person gets 0.5 litre a day from food — especially from fruit and vegetables. Another 0.5 litre comes from the chemical action of respiration in the cells of the body (Chapter 7). So we need to drink just a little more than one litre of water, tea, milk or other beverage each day.

fig 10.2 *Many small animals, such as this field mouse, do not normally drink water. They get all the water they need from their food and from the water that is produced by respiration in their cells. They sweat little (only on the soles of their feet), their urine is very concentrated and their droppings (faeces) are very dry, so they can live well on dry foods (grain) and in places where no liquid water is available.*

fig 10.1

During each day we lose two litres of water. Half of this is lost as urine. Almost half is lost as sweat and as the water vapour in the air we breathe out from our lungs. A small amount is contained in our faeces. The amount lost as urine can vary: if we drink a lot of water, we lose the excess amount by passing a larger volume of urine; if we get hot and sweat a lot, the amount of water in the urine is reduced. So the amount of water contained in the body stays constant.

The amount of water needed by the body is only a tiny fraction of the amount used for

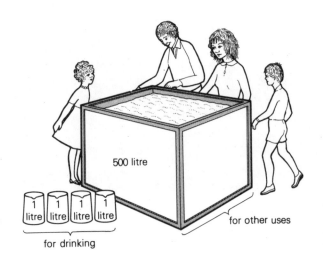

fig 10.3

other purposes. The average European family uses over five hundred litres of water each day.

Some of our technologies use large amounts of water:

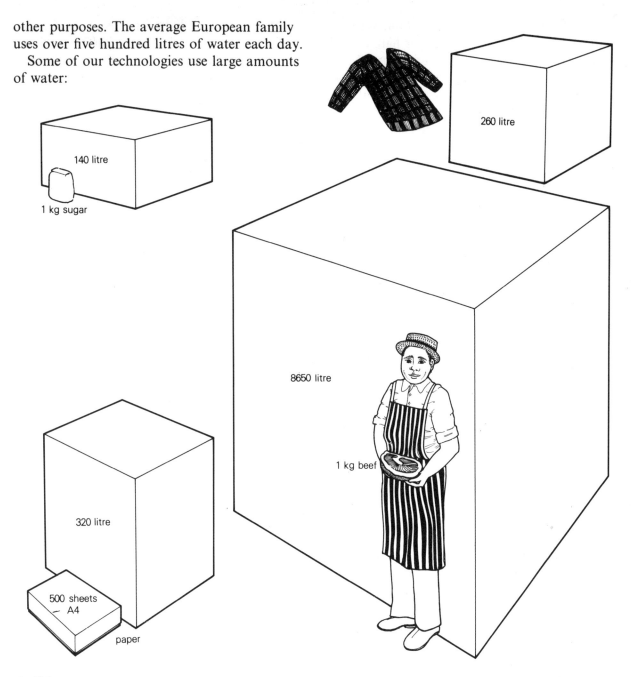

fig 10.4

It follows from the above that the technology of collecting, storing and distributing water is an important part of our present lifestyle.

Almost all (over 99 per cent) of the world's water is in the sea and polar ice caps. There are plans to collect icebergs and tow them to desert lands, and it is possible to extract pure water from sea water by distilling it, but these methods are relatively expensive and are not yet well developed. We get our water from rain, from rivers and lakes, and from underground. Figure 10.5 shows that *all* our water really comes from *rain* (or snow or hail, in winter). From where does the rain come? Most rain falls in hilly parts of the country (the west and north in Britain). Some of the rainwater soaks into the ground and may pass through the rocks for long distances. Then we can pump it up from deep wells and use it. The remaining rainwater collects in streams and rivers. Water can also be pumped from these. If taken for industrial use, most of the water is returned to the river again after it has been used. Before this is done, it may be necessary to heat it to remove harmful substances (Chapter 14). Similarly, water, after it has been taken and used domestically, is normally treated in a sewage works before it is discharged into the river or the sea. Rainwater and the water from mountain streams are normally pure and fit to drink but, in a country which is as crowded as ours, there is great danger that any water supply will become contaminated with disease organisms or harmful chemicals. For this reason the water which is supplied for use in our homes is first treated in a water works. Figure 10.7 shows the main stages in water treatment.

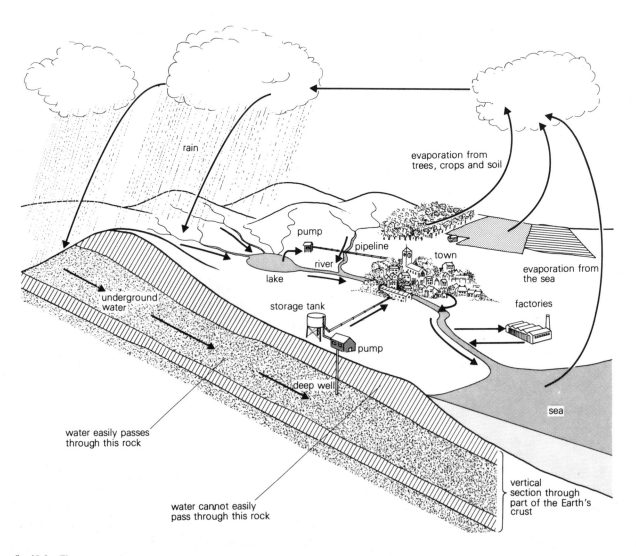

fig 10.5 *The water cycle*

fig 10.6 *This reservoir holds the water supply for the city of Birmingham. When there is no natural lake for water storage, a valley can be dammed to make an artificial lake.*

fig 10.8 *Filter beds. A thin layer of algae grows on top of the bed; many one-celled animals live there. The jelly-like layer of algae filters the finest particles and most bacteria from the water. Many of the bacteria become the food of the one-celled animals. For faster filtering, many modern systems use large, closed metal drums filled with sand and a chemically-made jelly.*

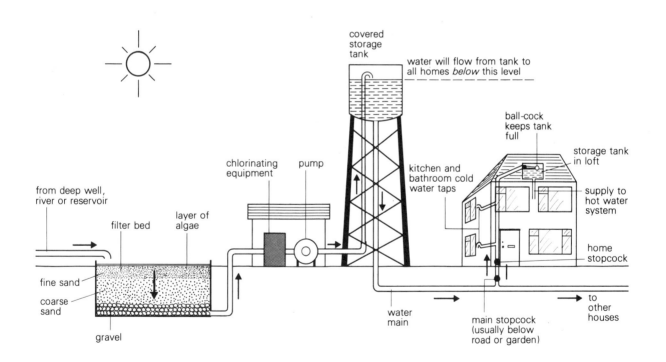

fig 10.7 *Treatment of water for use in the home. Water from rivers is held in a settling tank to allow large particles to settle out before the water goes to the filter bed. Chlorine is added to destroy any bacteria that are not removed by filtering. The water is stored in a high tower to give the necessary steady water pressure to drive it through the mains. In hilly districts, storage tanks are often placed at ground level on the highest hills instead of on towers.*

fig 10.9   Farming the desert. Rainfall is slight, but all that falls is carefully collected as it runs off the hillsides. Hardly a drop is wasted, making it possible to grow plentiful crops in a barren desert. Some water is stored in underground tanks for drinking. This technology, invented over 2000 years ago, is still the best for cultivating the desert.

Things to do

1   Study your local water supply. Find out the source of water, how it is treated and how it is stored and distributed. Collect photographs and draw maps to illustrate all the stages through which the water passes between the time it falls as rain and the time it comes from the tap in your home.

2   Try to imagine what your life would be like if you could have only three litres of water each day to use in your home. What effect would this have on the way you live? What would be the effect of such water shortage on a whole community? Do you know of any parts of the world where people regularly live under such conditions?

3   In the early part of this century, most people living in country areas of Britain obtained their drinking water from shallow wells. What are the dangers of using shallow wells? What can be done to this well water to make it safe to drink?

11 Our food

Like many other animals, early humans lived as hunters and gatherers. They spent all day in the forests, hunting for small mammals, birds and reptiles or gathering all the fruits, nuts and tubers that they could find. They also probably collected and ate fat grubs of insects, and the occasional honey combs made by wild bees.

To get enough food to keep themselves from starvation they needed to spend *all day* and *every day* hunting and gathering. There was little time left in which to do anything else.

But early humans, like people nowadays, had several advantages over the other animals:

1 The human lower arm can twist, so the hand can face up or down (Figure 11.1). We can position the hand to perform many different operations – lifting, pressing, turning and so on.

fig 11.1 *The ability of the lower arm to twist is useful in so many ways*

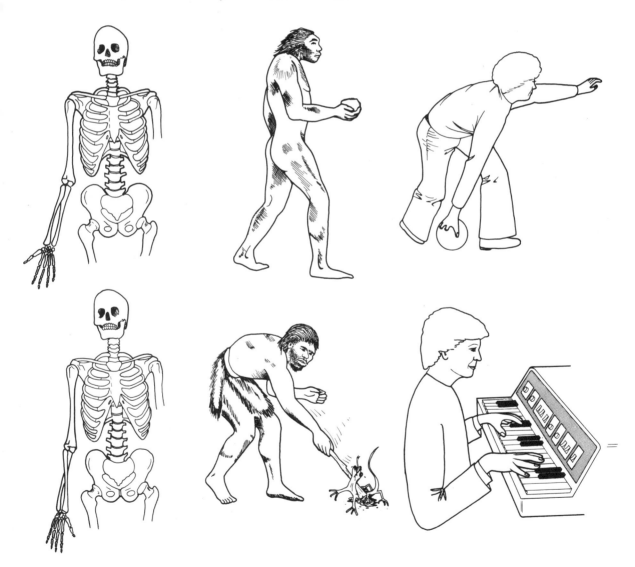

2 The human hand has great flexibility combined with strength (Figure 11.2). The twistable lower arm and the flexible hand make it easy to use all kinds of tools. In this sense, 'tools' include weapons for catching and killing animals to feed on.

fig 11.2 Because the human hand is so flexible, we can do many things that other animals cannot. The use of tools increases our power and ability in many ways.

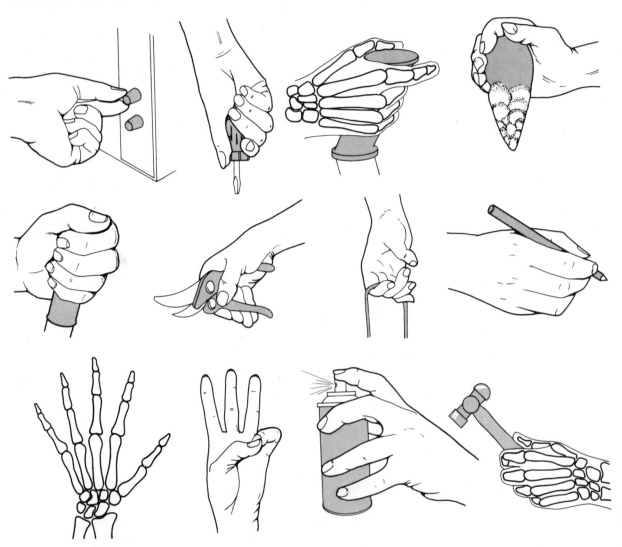

3 The human brain is the most complicated of all animal brains. This makes it easy for us to learn how to use tools and to invent new and better tools.

4 Humans are social animals. They have languages with which they can *communicate* with other humans. They can *teach* each other how to make and use tools, which are the best foods to eat and which dangers in life must be avoided. When a human thinks of a good idea, or invents something new, other people can easily be told about it. Humans can *cooperate* in performing tasks – for example, a band of hunters can work together to catch animals much larger and stronger than themselves.

Hunting by early humans was made more

and more effective by the invention of hunting tools, such as axes, spears, arrows and traps. These must have been the first examples of technology – perhaps the first was simply a heavy stone or stout stick held in the hand. Ways of making sharper tools were invented – for example, chipping flinty stones to give them a razor-sharp edge. Other tools such as the bow and arrow, the boomerang and the bolas made it possible to catch and kill an animal without the need to get close to it. By using these weapons, by cooperative hunting, by improving hunting techniques (for example, by digging pit-falls, by the discovery of poisons for arrows or by using hunting dogs) the food supply was increased enormously. Then the human population was able to increase, too.

While the hunting methods were being made more efficient, it was found that by clearing a small area of forest and sowing seeds there, crops could be grown. This saved hours of searching the forest for fruits and nuts. The

fig 11.4   A stone axe. This simple tool was made a few years ago in New Guinea, where such stone tools were widely used until recently.

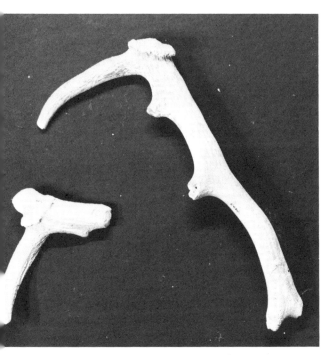

fig 11.3   A flint-miner's pick made from a deer antler. The point was probably hammered into natural cracks in the rock, to lever out blocks of chalk or flint. This was probably made and used about 4000 years ago

discovery of fire helped, too, for many vegetable foods that are too tough to eat when raw are much more eatable when cooked. The human diet could now include many kinds of tuber and root as well as leafy vegetables. Gradually the ideas of agriculture were developed. Crops were grown and the types and varieties that gave the best results were discovered. In a similar way, domestication of some of the animals began, too – goat, pig, sheep, cattle and horse. Grasses and other crops were grown for feeding to farm animals.

Agricultural technology began over five thousand years ago. It had two important effects:

1   People had to settle in one area for a year or two at least, so as to be able to harvest their crops. They no longer wandered continuously through the forests.

2   The early farmers were easily able to produce far more food than they needed for themselves. No longer did everybody have to spend every day looking for food. While the farmers

fig 11.5 *Irrigation was an early method used to improve soil fertility. Above, we see a shaduf being used in early Egypt. Modern methods include lawn sprinklers and complex sprayers for agricultural crops, but the shaduf is still in use, as seen below in a photograph taken a few years ago beside the Nile in Sudan.*

fig 11.6 *Today, few people live as hunters and gatherers. The chief exceptions are the fishermen. Although fish are cultivated in tanks (fish farms) in some parts of the world, the greatest numbers are still caught by gathering them from the open seas.*

produced food, other people had time to do other things. They were free to invent and improve new technologies. New trades arose — people worked as potters, weavers and toolmakers. In return for their produce, they obtained food from the farmers.

Today, in the industrial countries, relatively few people work on the land, producing food. Most people work in manufacturing industries, in service industries, or in government, and can buy the food they need with the money they earn. In this way our lives today are very different from the lives of our ancestors.

Things to do

1   What are the most important inventions and discoveries that have improved agriculture during the last five thousand years?

2   Name some manufacturing industries. Name some service industries. What percentage of the parents of people in your class work in (a) manufacturing industries, (b) service industries, (c) government, (d) agriculture and (e) hunting and gathering?

12 Protecting our food supplies

Food needs protecting at all stages in its production. If we do not protect it, we lose it to some other animal, or to the decomposers. Then there is less for us to eat; we may starve. Livestock need protection from carnivorous animals who, like us, regard a chicken or a young lamb as a satisfying meal. The poultry-keeper makes a strong fence around the chicken-run to keep out foxes. In many countries, herdsmen watch the sheep and cattle all day and night to protect them from attack by wolves or lions or hyaenas (and also protect them from being stolen by other hungry people!). Livestock need protection from disease caused by bacteria, roundworms and other organisms since a sick animal is not a source of food that is fit to eat. Veterinary science helps here. Our crops need protecting too, particularly from being eaten by insects, from fungal diseases and from other organisms that attack the plants. Insects eat about one third of the food we produce, so if we can prevent insects from eating food when it is in the field or while it is in storage, great savings can be made.

Much of our food production is seasonal. When crops ripen, there is more food than we need, but there will be no new supplies until another twelve months have passed. Most of the harvest must be preserved, so that we may have a supply of food all the year round and so that the newly-harvested supplies do not rot and spoil before we have been able to eat them. Some foods, such as cereals and cereal flours, potatoes, carrots and other root crops, can be stored in a cool place for many months. We have to protect them from rats and mice and from certain insects such as grain weevils and flour beetles. Other foods do not keep in eatable condition for more than a few weeks or days. For these we use one of the special technologies of food preservation:

1 **Drying**: the oldest known method of food preserving. Many kinds of fruit are preserved by putting them in the sunshine until they are dry. Can you think of examples? Can you think of other foods that are preserved by drying?

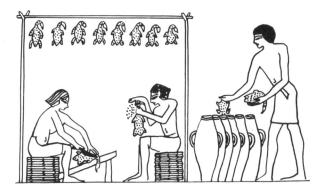

fig 12.1 Several thousands of years ago the Egyptians used salt for preserving game birds

fig 12.2 Food canning – a modern technique requiring complicated factory machinery

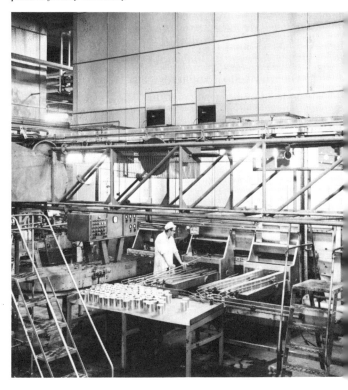

fig 12.3 Smoking of herrings to make kippers is an ancient craft requiring skill and knowledge but only simple equipment

2 **Smoking**: the food is hung over wood fires; this dries them but also helps kill bacteria. Name some smoked foods.

3 **Pickling** by adding vinegar or by fermenting to produce alcohol. Some examples are given in Chapter 7.

4 **Salting**: the food is kept soaked in a strong solution of salt; meat and runner beans are often preserved in this way.

5 **Jam** is made by adding a lot of sugar to boiled fruit. Bacteria and fungi cannot grow if the amount of sugar present is very great.

6 **Cooking** helps preserve food for a short time.

7 **Canning**: the food is sealed in airtight cans, after some kind of heating to kill any bacteria present. Name some foods that are often bought in cans.

8 **Bottling** in a solution of sodium meta-bisulphite (Campden tablets) is often used for preserving fruits and vegetables at home.

9 **Pasteurisation** is used for treating milk, ice-cream, liquid egg, beer and vinegar. It does not kill all the organisms present, but kills all those likely to cause disease and kills enough of the others to prevent them from spoiling the food quickly.

10 **Refrigeration and deep-freezing**: one of the most modern methods. Name foods that are commonly preserved in this way. What are the advantages of this method of preservation? What are the difficulties?

Things to do

1 Collect pictures and samples of preserved food to illustrate all the different methods of food preservation listed above.

2 Look again at the conditions needed for a healthy active life (Chapter 1). For each of the preserving methods given above, say how we make the conditions in the preserved food *un*suitable for decomposing organisms.

3 What diseases and pests attack and spoil the crops grown in your garden? Make a list of these and say what the gardener can do to prevent these attacks or make them less serious.

13 Keeping out the cold

Birds like sparrows live outdoors all through the winter. To keep warm, they fluff out their feathers, just as the bird in the photograph is doing.

fig 13.2 Bees build in any convenient place (including bee-hives) that is dry and sheltered from the cold. There they are safe from wind and rain. They defend their living space from enemies, such as wasps, which may attack them. In winter they crowd close together, so keeping themselves as warm as possible.

fig 13.1 A fieldfare in winter

When it is very cold they stay in a place away from the wind, the snow and the coldness – perhaps a hole in a tree, or in a dense covering of ivy on the wall of a house. Most animals need shelter in winter – unless they live in a warm country or, like many of the birds, are able to migrate to a warmer country in wintertime. Yet, even in hot countries, it can be cold at night. In the deserts of Africa, the temperature may reach 40°C during the day, yet fall below zero at night.

Animals need to protect themselves against temperatures that are too high and those that are too low. They need to protect themselves against wind and heavy rain. Wind and rain make life difficult and cause damage. A wet animal becomes very cold when exposed to wind. If you have ever been caught in a rainstorm and have then had to wait for a bus on a windy corner, you will know just how cold you can become.

For a small animal (such as an insect), it is usually easy to find a place to shelter in. It can crawl under a stone or find a small crack in the bark of a tree. There it is safe from the worst of the weather. It is also safe from its enemies, for it is out of sight and hard to find. Bigger animals often cannot find shelter so easily, but they may solve the problem by building their own shelter. In fact, many animals build or make a shelter. They bore holes in bark, they cut holes in the wood of trees, they build nests, they build tubes – can you think of the kinds of animals that make shelters?

Like the sparrow, humans are warm-blooded animals. Like the sparrow, they need a shelter from the coldness of winter. Early humans probably sheltered in caves. This was a good solution to the problem of winter, when there were not many people on Earth, but, there are few caves in the world, so today it would be impossible for us all to find shelter in this way.

fig 13.3 (above)   This park seems to have few animals in it in winter, yet there must be many there, if we could only find them, waiting for the warmer days of spring. Where, in the area shown in this photograph, would you expect to find small animals sheltering from the winter cold? What kinds of animals would you expect to find? If you have a park or garden that you can visit and explore, you could try to find the places in which animals shelter – but do not disturb the animals, for they will probably die if forced out into the cold.

fig 13.4 (below)   A piece of bark from a dead elm tree has been peeled off; looking at it from the inside, we can see the tunnels bored by the elm bark beetle. This beetle finds shelter there. The vertical tunnels are cut by the female; she lays eggs along the sides of a tunnel. Then the larvae hatch and bore the smaller tunnels that radiate from the large ones; they feed on the bark as they go. Later they pupate in their tunnels. Hatching as adults, they bore their way out and fly to another tree. Unfortunately, they carry a disease fungus from infected trees to healthy trees – the fungus that causes Dutch elm disease. The disease is fatal to trees and many elms in southern England have died from it in recent years.

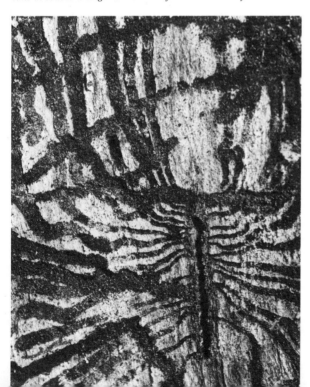

Instead, we have invented ways of building shelter – anything from a few large leaves laid on a frame of twigs, to a multi-storey tower of centrally-heated flats. Now, large numbers of us can live in areas where there are no caves or other natural shelters. For thousands of years we have been thinking of new and better ways of building houses, and we have discovered many new building materials. This is another of our activities as a technological animal. House-building is an important part of our lifestyle. You can find out more about this in Chapter 14.

We also protect ourselves from cold by wearing clothes. Our ancestors needed to leave their caves in winter to hunt for food, so they used the skins of animals as clothes. Later the technologies of spinning, weaving and knitting were invented. We have discovered many natural fibres that can be used for clothing. In this century we have found out how to make fibres artificially and how to make clothes cheaply from these fibres. We need no longer worry about keeping out the winter cold.

Things to do

1   Find out why
(a) fluffing out the feathers helps keep the bird warm,
(b) deserts can get very cold at night and
(c) wind and wetness *together* cause cooling.

2   Make a list of the animals that spend the winter in your part of the country. Where do they go when the weather is cold? What do they feed on in the winter? Which ones hibernate for most of the winter?

3   Which animals leave your area in autumn to migrate to warmer parts of the world? To which countries do they go? Which animals come to your area from colder areas further north?

4   Which fabrics are the best for keeping out the cold? Design an experiment to measure how well a fabric acts as a heat insulator, and test several fabrics to find the best.

fig 13.5 Clothes are used for many reasons, such as keeping warm, keeping cool, protection from harmful conditions, decoration and ceremony. Which clothes in the drawing are worn for which of these purposes? (Some may be for more than one purpose.) In the case of protective clothing, say what harmful condition it protects us from.

14   A house

A house is a shelter   A house protects us from temperatures that are too low and from temperatures that are too high. It also protects us from some harmful conditions of the world outdoors:

fig 14.2   People who live near busy airports need efficient sound-proofing in their houses

fig 14.1   This is a village in the Highlands of New Guinea. During the day, the outdoor temperature is comfortable, but it can be very cold at night. Most of the houses are round, with no windows and only a small doorway. At night, a wood fire is lit inside; the smoke finds its way out through the thatch of the roof. In what ways are these houses suited to the local climate?

Water: rain and snow come from above; flood water and dampness from the soil below. A good house keeps all these out. Dampness is bad for health and it spoils the things that we keep in our house, such as food, clothing and furniture.

Wind: it causes damage; it makes us cold.

Noise: from traffic, jet aeroplanes, railways, road-drills. These make loud noises that are very unpleasant. Even quieter noises, such as the sound of a radio playing next door, can be very annoying at times. A good house keeps out most of the unwanted noise.

Enemies: we are often warned to lock our doors and to have secure fastenings on our windows to keep out thieves. In hot countries people usually like to keep their windows wide open, day and night. They need bars across their windows to stop thieves from climbing in. They need a screen of fine netting to stop insects from flying in. Mosquitos may bring malaria (Chapter 17) or other diseases. Houseflies must be kept out, and so must all animals which may get in to the house and attack us or spoil our food or possessions. What animals does your house give you protection against?

A house is our environment   We control the environment inside the house, to make it as comfortable to live in and as pleasing to look at as we can afford. We can have heaters to make it comfortably warm, or an air-conditioner to make it cool in hot seasons. If the windows are small or if there are no windows at all, it is easier to keep the house warm (or cool, in a hot climate), but then we need lamps to see by. There is also the problem of ventilation. A good house must allow enough fresh air to enter – but not a freezing gale!

A house is a home   We are **social animals** and most people are happiest when living as a member of a family. A house is a base for a

fig 14.4 This is the kind of house built on the hot, humid coasts of New Guinea. The air is damp and warm, but breezes blow in from the sea daily. Compare this with the Highland houses in Figure 14.1 and explain the differences. What are the advantages of building the house on stilts?

fig 14.3 Modern houses are much better than older houses at keeping out the cold and keeping in the heat. The heaters used nowadays are better at heating the room and its occupants. The result is that people can keep warm while wearing less than in the old days. Present-day fashions can be light in weight and comfortable to wear

family: it is a home. A home is much more than a living-place made of bricks or wood or cloth or blocks of snow. Any place – from a cave to a luxurious pent-house apartment – can be a home *if* the people living there act to-

gether to *make it* a home. We are **territorial animals** too. We need to have a place that we feel belongs to us, a place where we feel we have the right to be and to do as we please. Our home is our territory, where we feel secure.

Things to do

1   Collect photographs or drawings of houses from all parts of the world. Include any kind of dwelling such as tents, igloos and caves. Make a display of the pictures in your classroom. Beside each picture pin a small card with notes explaining
(a) how it keeps out the cold (or the heat, if it is a dwelling found in a hot country),
(b) how it keeps out the rain,
(c) how it keeps out enemies and harmful animals,
(d) how it keeps out noise (if it does), and
(e) which cheap local building materials are used for building it.
You could also make cardboard models to show what some of the dwellings look like.
2   Make a display of drawings, photographs and manufacturer's leaflets to illustrate modern building methods and materials – double-glazing, roof insulation, cavity walls, wall

insulation, damp courses, gutters and drains (to keep the house dry and prevent the land around from becoming flooded). If possible, include samples of the materials used.

3 Find out as much as you can about the survival methods used by mountaineers and by members of the armed forces.

4 Suppose that you were asked to design Moonbase Alpha. The Moon has no atmosphere, so there is no oxygen, no rain and no wind. During the Moon's day, the temperature at the surface rises to about 100°C; during the night it falls to −180°C. What are the problems of making the environment inside Moonbase Alpha suitable for people to live there? How would you try to solve these problems?

15   Getting rid of waste

vities of the decomposers thus keep the natural environment free from large accumulations of waste materials.

Much of the waste produced by humans can be disposed of in the same ways. When there were few people on Earth, there were no pro-

All organisms produce waste materials:

1   **Gaseous waste**: for example, carbon dioxide from respiration, oxygen from photosynthesis

2   **Solid waste**: for example, undigested materials from the gut

3   **Liquid waste**: from various processes in the body. This usually consists of water with dissolved waste substances, for example, urine

4   **Dead bodies or dead parts of bodies**: for example, dead leaves falling in autumn, feathers and hairs that fall out.

Gaseous waste escapes into the atmosphere and is recycled (Chapter 33). Organisms living in water usually pass these gases into the water, in solution, from where they may pass into the atmosphere.

The useful materials in other wastes are recycled, mainly by the decomposers (Chapter 8). In natural conditions nothing is wasted. All waste is used over and over again. The acti-

fig 15.2   *The sprinkler at a sewage plant*

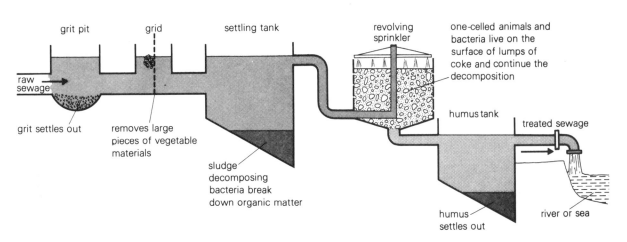

fig 15.1   *One method of treating sewage. The sludge tank is deep, for at that stage the decomposing bacteria are the kind which live best when there is no oxygen. At the sprinkler stage, the bacteria and one-celled animals need plenty of oxygen, which is freely* *available in the tank of loosely-packed coke. The sludge and humus are later dug out from the tanks, dried, and then used as organic fertilisers on the land.*

blems. But today, when there are large numbers of people living close together in towns and cities, waste disposal is a serious problem. At the sewage works, decomposers are used to treat raw sewage and make it suitable for discharging into rivers or the sea. If raw sewage is allowed to go directly into the rivers, it may spread disease. The large quantities of organic material in raw sewage provide a rich source of food for the decomposing bacteria in the river. They multiply at a high rate and so *use up oxygen* at a high rate. The amount of oxygen in the water is decreased to such an extent that fish and other animals are not able to live there. Sometimes, raw sewage or partly-treated sewage has been allowed into rivers; when this has happened, the natural population of the river has been very badly affected and it has taken years to recover. Although properly treated sewage may be run into rivers without harm, too much of it upsets the water life. Treated sewage contains large amounts of mineral salts (produced by decomposition) and these cause excessive growth of some kinds of algae, so making conditions unsuitable for many other kinds of organism. The balance between the different kinds of plants and animals in a natural environment is easily upset by major alterations in their living conditions – and the change is usually for the worse.

Unfortunately, many of our waste materials are not *natural* materials – they cannot be decomposed by the activities of the decomposers. The decomposers simply do not have the enzymes needed to break down the molecules of, for example, a plastics ice-cream carton. Although a food-can is corroded by *chemical* action, this is usually a very slow process. Much of this refuse accumulates around us year by year, in every-increasing amounts.

Things to do

1   Bury small scraps of different materials about 20 cm below the soil surface. Use leaves, vegetable-peelings, slices of potato, paper,

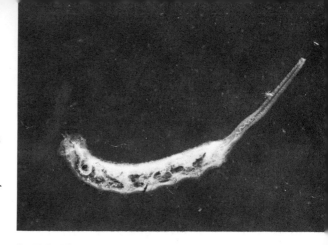

fig 15.3   These animals can live in water that contains little oxygen. The rat-tailed maggot (above) has a tube by which it breathes directly from the atmosphere. The bloodworm (below) has the red substance, haemoglobin, in its blood; this substance absorbs oxygen very readily, so the bloodworm can obtain enough oxygen, even when little of it is available.

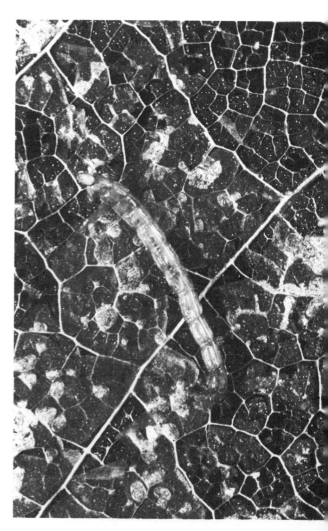

scraps of wood, scraps of various plastics, pieces of fabric, glass, small metal objects and anything else you can think of. Keep the soil moist in dry weather. After four or five months, dig up the pieces and look for signs of their being eaten by animals or being decayed by decomposers.

2 Can we afford to waste so much of the materials that we produce? What methods can be used for recycling used materials? What can you do in the home and garden to recycle your refuse and to reduce the amount of refuse you produce? What is being done by your local government authority or by local industries?

fig 15.4 The pictures on p. 53–54 illustrate some of the problems created by technology – what are the problems? What solutions are there?

(above) Refuse disposal costs money – the owner of this refuse would not meet the cost; the community suffers
(below) Refuse from a tanker, or spillage from a wrecked tanker, can cause widespread damage to wild life and make beaches unpleasant for the community

(above) Many kinds of detergent are not decomposed by bacteria. By using 'biodegradable' detergents (detergents that can be decomposed by bacteria) we can prevent detergents having a harmful effect on river life.
(below) Refuse that cannot be recycled by the decomposers

What goes up must come down! Where?

(above) The exhaust gases from car engines contain carbon monoxide and compounds of lead – both highly poisonous substances

(above) Pesticides and herbicides are highly poisonous, and many are not decomposed after they have done their job. We must make sure that they are all used in a way that does not damage the natural environment or cause danger to the community.

(above) Radioactive waste materials continue to be dangerously radioactive for thousands of years. Safe disposal of these wastes is a problem yet to be satisfactorily solved.

(above) Artificial fertilisers are widely used today. Water draining from fields is rich in minerals, which enter our rivers. In the rivers, this excess mineral material favours the growth of algae at the expense of the growth of other organisms. The mineral material often includes soluble nitrates which, in sufficient quantity, are harmful when drunk by babies.

(left) In the past, solid wastes from cattle were recycled by being spread on the land as manure. Nowadays, many farmers cannot, or will not, use this (why?). It must be disposed of through sewage systems for, if this is not done, our rivers will become polluted.

16 Insects

Insects outnumber all the other kinds of animal on Earth put together. They are the most successful kind of animal we know. Some live on land, others live in water, many fly in the air. They have adopted several lifestyles – some are herbivores, some are carnivores, some are parasites (Chapter 18). Can you think of examples of each?

An insect has these features:

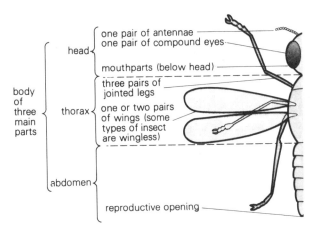

fig 16.1

fig 16.2 (below)   The antennae (or feelers) carry smell-detecting organs. Compare the antennae of this pearl-bordered fritillary butterfly with the antennae of other kinds of insects.

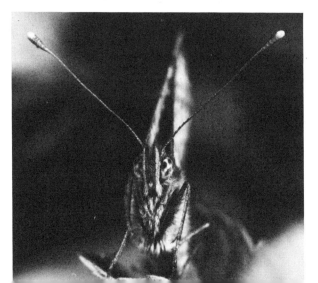

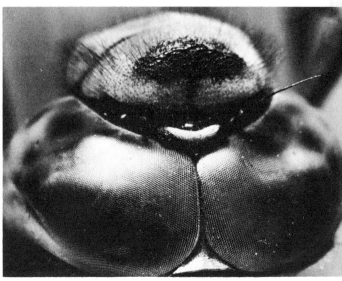

fig 16.3   The large compound eyes of this hawker dragonfly consist of thousands of eye units, each with its own tiny lens

The hard outer skeleton is made of **chitin**. This protects the body of the insect from damage and helps protect it from being eaten by other animals. To allow the insect to move, the plates of chitin are joined together by a thin membrane of a more flexible material. The muscles of the insect are attached to the inside of its skeleton.

fig 16.4   The lower part of a housefly's leg. Note the joints. The bristles are part of organs that detect smells and touch. What use are the claws? What use is the sticky pad between them?

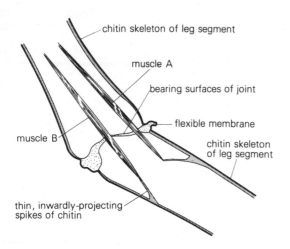

fig 16.5 A sectional view (left) of the joint between two segments of the leg of an insect. When muscle A contracts, the leg is made to bend at the joint. When muscle B contracts, the leg is made straight again. Our own limbs work in a similar way (right). Which of the two muscles shown is the one that makes our arm bend? Which makes it straighten?

Although most insects have the features described on p. 55, there are several main families that are easy to recognise:

fig 16.6

(a) Butterflies and moths: large wings covered with tiny coloured scales. In most butterflies (e.g. this swallowtail) and many moths the scales make brightly-coloured patterns.

(b) Beetles: hard front wings protect the transparent rear wings, used for flying

(c) Flies: the rear wings are modified as a pair of club-shaped haltres, *which help the fly to keep its balance while flying*

(f) (above)   Bugs: the water scorpion belongs to a group of bugs in which the front half of each fore-wing is hard and not transparent. Like all bugs they feed by sucking juices. The water scorpion sucks blood and body fluids from the small pond animals it catches with its large fore-legs.

(g) (below)   Earwig: front wings are hard and short, protecting the delicate hind wings folded beneath them

(d) (above)   Bees and wasps: two pairs of transparent wings, the front pair larger. The ants belong to this group, but worker ants have no wings.

(e) (below)   Bugs: aphids (e.g. greenfly) hatched in spring have transparent wings and fly to the crops on which they feed, sucking the juices from stems and leaves. Aphids produced in summer are wingless.

The mouthparts of insects are complicated. In different kinds of insect they are modified in various ways to suit the food the insect normally eats.

fig 16.7   *The cockroach has strong jaws with toothed edges It eats almost anything.*

fig 16.9   *Mosquitoes have a tube-like mouth with a sharp tip for piercing and sucking blood (× 50)*

fig 16.8   *Butterflies have a tube-like mouth for sucking nectar from flowers. It is curled up when not in use.*

fig 16.10   *The housefly's mouth has a flattened pad at the tip. It spreads saliva over the food and then sucks up the liquefied food.*

The life stories of several kinds of insects are discussed in Chapter 22.

Things to do

1 Find out which are the five most common insects in your district. If you were asked to do this again next year, do you think you would necessarily give the same answer?

2 Study the insects that live in ponds, or rivers or can be kept in aquaria in the school laboratory. Study the different ways in which aquatic insects swim. Study the different ways in which they obtain oxygen.

3 Study the insects visiting flowers in the garden. Find out why they come to the flowers (for nectar or for pollen?). Find out how they help pollinate the flowers.

17 Disease

The main ways that disease organisms can enter our bodies are shown on p. 31. Once inside, they multiply there, causing disease. For the disease to spread to another person, some of the disease organisms must *leave* the body and then *be carried* to the body of the other person. There are many different ways in which this can happen, depending on the type of disease organisms and on the *habits of the people* concerned. We cannot alter the disease organism, but we may be able to alter our habits to make it more difficult for the disease to spread.

The organisms may leave the body by the way they came in – for example, through the mouth or nose. Often, the symptoms of the disease help them to leave; coughing, sneezing and a running nose help scatter organisms into the air or spread them onto hands, furniture or food, from where they are soon carried to other people. What can be done to reduce these ways by which disease spreads? Organisms that live in the gut generally leave the body with the solid waste (faeces). With good hygiene (Chapter 19) and with proper treatment of sewage, there is little risk of the organisms being spread to other people.

When a person has been made sick by a disease brought by an insect, we naturally wish to help that person to recover. Medicines are available for treating many of the diseases. But to *prevent* disease is better than to *cure* it. If the insect is the *only way* in which a disease is spread from person to person, we should try to eliminate that kind of insect. By eliminating the insect carrier we can prevent the disease from spreading and eliminate it too.

fig 17.2   The tsetse fly lives in parts of Africa and feeds on human blood. It carries the trypanosomes of sleeping sickness from person to person.

fig 17.3   Different kinds of mosquito carry different kinds of disease. This kind carries the virus that causes yellow fever.

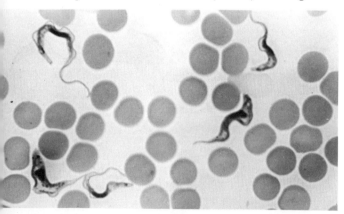

fig 17.1   The blood of a person who has sleeping sickness. The sleeping sickness parasites are one-celled animals (trypanosomes), seen here among the red blood cells (rounded) of their host. (×1200)

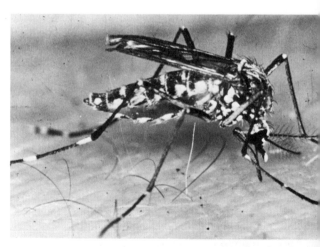

The disease MALARIA A severe fever, often fatal

The cause

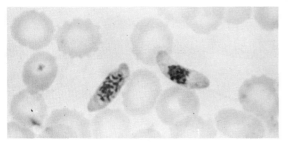

PLASMODIUM a one-celled animal parasite. It lives inside red blood cells and in liver cells.

The carrier

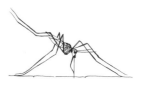

The female ANOPHELINE MOSQUITO, which has piercing and sucking mouthparts

How the disease is spread

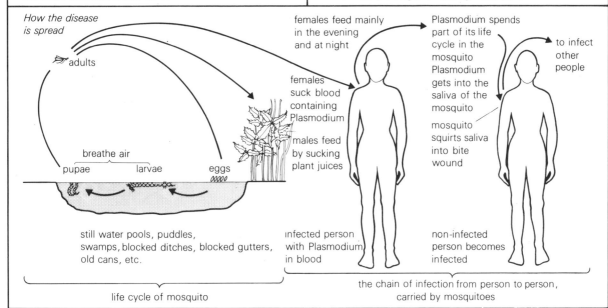

females feed mainly in the evening and at night

adults

females suck blood containing Plasmodium

males feed by sucking plant juices

breathe air

pupae   larvae   eggs

still water pools, puddles, swamps, blocked ditches, blocked gutters, old cans, etc.

infected person with Plasmodium in blood

Plasmodium spends part of its life cycle in the mosquito Plasmodium gets into the saliva of the mosquito

mosquito squirts saliva into bite wound

to infect other people

non-infected person becomes infected

life cycle of mosquito

the chain of infection from person to person, carried by mosquitoes

How the spread of the disease is prevented or reduced
- drain swamps
- spray pools with oil
- spray pools with insecticide
- clear away litter, such as cans
- spray homes with insecticide
- screen windows and doors of houses with fine netting
- wear long-sleeved clothes and long trousers in the evening
- infected persons should take anti-malarial drugs to kill Plasmodium in the blood
- non-infected persons in malarial areas should take anti-malarial drugs

fig 17.4

The disease DYSENTERY and other infections from contaminated food

The cause

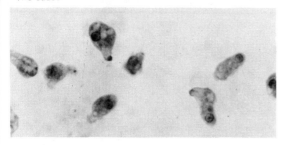

Many kinds of bacteria may be involved, including the dysentery bacillus (above)

The carrier

The HOUSEFLY carries bacteria on its hairy body and legs, its sticky foot pads, in its saliva (which it squirts on food) and in its faeces (which it leaves on food)

How the disease is spread

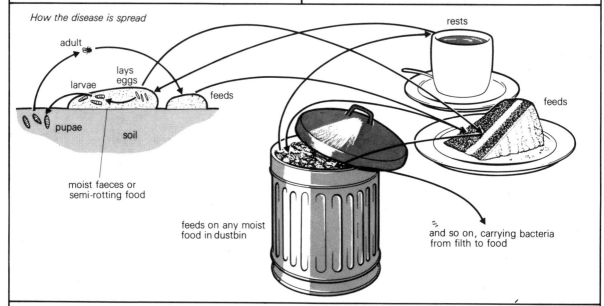

adult
lays eggs
larvae
feeds
pupae
soil
rests
feeds
moist faeces or semi-rotting food
feeds on any moist food in dustbin
and so on, carrying bacteria from filth to food

How the spread of the disease is prevented or reduced

- dispose of faeces by WC and sewage treatment

- make sure that waste food is put in a covered dustbin

- use wire screens on larder windows

- use dangling plastic strips on doorways in summer

- keep food covered or in cupboard or refrigerator

- keep crockery and cutlery in cupboards

- use insecticide or fly-papers in the home (especially in the kitchen) and food shops

- food sold in shops must be kept covered

fig 17.5

The disease PLAGUE (=Bubonic Plague = Black Death)

The cause

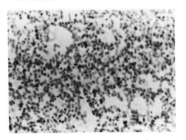

the bacterium that causes plague

The reservoir

brown rats and black rats have the bacterium in their blood, but are resistant to the disease

The carriers

rat flea and human flea become carriers when they feed on infected blood

How the disease is spread

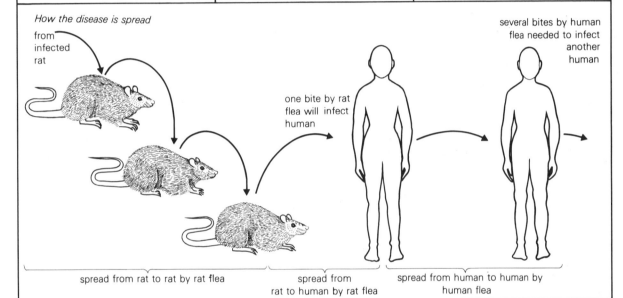

from infected rat

one bite by rat flea will infect human

several bites by human flea needed to infect another human

spread from rat to rat by rat flea

spread from rat to human by rat flea

spread from human to human by human flea

How the spread of the disease is prevented or reduced

- make homes rat-proof
- make food stores (e.g. grain stores) rat-proof
- make sewers rat-proof
- use rat-guards on the mooring ropes of ships
- use insecticides in houses
- change underclothes often
- wash hair often

● use rat poison   ● take baths often

fig 17.6

Ronald Ross

Ronald ROSS, a Captain in the British Army, lived from 1857 to 1932. He served in India, where he collected information about the unpleasant and often fatal disease, malaria. This disease was extremely common in tropical countries and has in the past occurred even in Britain. The disease occurs in marshy areas where mosquitos breed, and it was thought that *Plasmodium*, the one-celled animal parasite that causes malaria, might be carried from person to person by mosquitoes. After working for months dissecting mosquitoes of all kinds. Ross discovered *Plasmodium* in the stomach of a mosquito of a certain kind. By further investigation he worked out the life story of the parasite and showed that it is transmitted to man *only by the mosquito*. He demonstrated that the best way to eliminate malaria – at one time the most killing disease in the world – is to prevent the mosquito from breeding. For this work, Sir Ronald Ross received the Nobel Prize in 1902. Since then there have been increasing efforts to put his ideas into practice. Gradually, by draining swamps and by large-scale use of insecticide, the malaria-carrying mosquito has been eradicated from many countries. This has been the result of a world-wide campaign launched by the World Health Organisation, an agency of the United Nations Organisation, in cooperation with the governments of the countries concerned.

Things to do

1   In each chart on pages 61–64 there is a list of ways of preventing the spread of the disease. Copy the items of this list, and for each one say exactly *why* it helps prevent the spread of the disease.

2   Find out about the epidemics of Plague that occurred in Europe in the fourteenth and fifteenth centuries, especially the Plague of 1665 in London and other parts of Britain. Do you think that such a plague could occur in Britain today?

18 Parasites

Disease bacteria, and several protozoa such as *Trichonympha* (Chapter 4), *Trypanosoma* (Figure 17.1) and *Plasmodium* (Figure 17.4), are examples of organisms that live *inside* the body of another organism. They get their food from that organism. They gain protection by being inside the organism's body. *Trichonympha helps* the termite inside which it lives, by digesting pieces of wood that the termite has eaten but cannot digest for itself. So *Trichonympha* and the termite live in partnership. Each helps the other. But disease bacteria and other disease-causing organisms (such as *Trypanosoma* and *Plasmodium*) actually *harm* the organism they are living in. They give nothing in return for the food and protection they gain. They rob the body of food and then may produce poisonous substances, or may damage the body of the organism – they may even cause it to die. Organisms that live inside another organism, getting food and shelter, and harming the organism they live in, are called **parasites**. The organism inside which they live is called their **host**. The lifestyle of the parasite is to gain food and shelter and give nothing in return.

Another kind of parasite lives *outside* the body of its host. It depends entirely on its host for food, but it may move from one host to another of the same or similar kind a few times during its life. The rat flea and human flea, shown in Figure 17.6, are examples of parasites of this kind. They spend all their life among the hairs on the skin of the host and suck a meal of blood whenever they need it. By living outside the body, they gain less protection than the parasites that live inside, but they are at least kept in a region of suitable temperature, near to the warm body of their host. Like all other parasites, they cause nothing but harm to their host.

The parasitic lifestyle is a highly successful one, as the examples below illustrate. All these parasites are adapted in various ways to their special type of lifestyle.

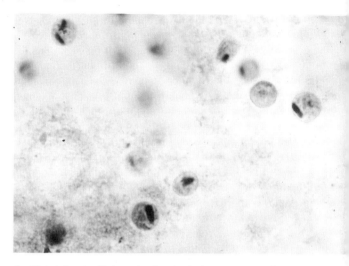

fig 18.1 (above) Entamoeba *lives in the human gut, causing amoebic dysentery. Like its relative,* Amoeba *(Chapter 4), it is a one-celled animal or protozoan. Unlike its relative, it does not have a contractile vacuole – why might it not need one?*

fig 18.2 (below) *The filaria worm, a kind of roundworm, lives in the tissues of the body: the tissues swell, producing the disease seen here – elephantiasis. The young filaria worms are carried from person to person by a mosquito.*

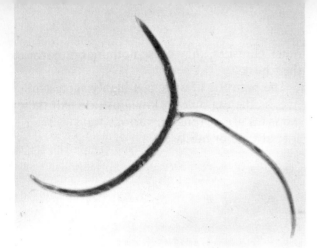

fig 18.3 *The hookworm (yet another kind of roundworm) lives in the gut of man. It attaches itself to the lining of the intestine, sucking blood and tissue fluids.*

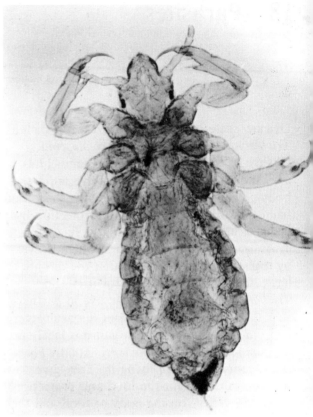

fig 18.5 *The body louse has hooked feet, that help it to hang on to the hairs of the body. It feeds on blood. Its flattened body is just the right shape for an animal that lives in the narrow space between skin and clothes. The louse carries the organisms that cause typhus and relapsing fever – two very serious diseases.*

fig 18.6 *Viruses are the smallest known living organisms. This is the virus that causes smallpox. Like all viruses, it is active only when it is inside the cells of its host: here it is growing on the membrane of a hen's egg (seen as dark spots). It relies on its host for almost everything; not only does it get food and protection, it even uses the reproductive mechanism of the host's cells to make new virus material. All viruses are parasites and almost all cause disease. Other diseases caused by viruses include measles, mumps, yellow fever, influenza and the common cold.*

fig 18.4 *The hyphae of this parasitic fungus penetrate the cells of the tree trunk and absorb soluble food materials. The 'bracket' shown here is the spore-forming part of the fungus. It produces millions of spores, which are blown away, possibly to infect other trees.*

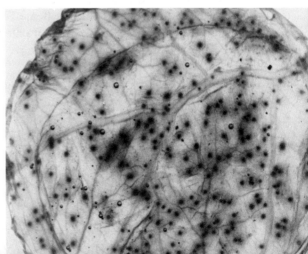

66

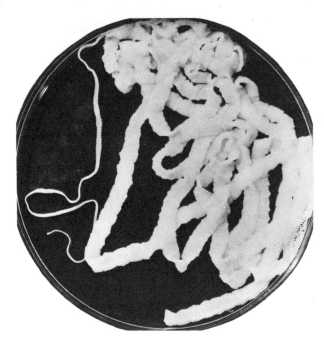

fig 18.9   This potato plant is infected by the parasitic fungus that causes patato blight. The leaves die and rot away, and the crop is very poor. This parasite caused severe famines in Ireland during the nineteenth century—find out more about these famines and their effects.

fig 18.7   This tapeworm needs no digestive system. It lives inside the human intestine and there it is bathed in fluid containing dissolved food materials. It simply absorbs through its surface the materials it needs.

fig 18.8   Mistletoe is partly a parasite. It takes water and minerals from the tree it lives on, but it has green leaves, so it can photosynthesise its own organic food materials. How are the seeds of mistletoe carried to another tree?

fig 18.10   The toothwort is a parasitic plant that grows attached to the roots of trees. Its small leaves have no chlorophyll, so it cannot photosynthesise for itself. It takes water, mineral salts and organic materials from its host. The only things it does for itself are to flower and produce fruits and seeds.

Things to do

1   Explain how *Entamoeba* can obtain the favourable conditions it needs (Chapter 1) while living in the gut of a human.

2   What parasites are commonly found on pet animals? What can be done to prevent their occurrence?

19 The fight against parasites

Parasites cause us nothing but harm. They harm our bodies and the bodies of our pets and our farm animals. They harm our crops. Many methods of combating parasites have been invented. Some of these methods have been described in earlier chapters. Here, th[...] more methods of special use against bacteria and viruses are described:

1 **Antiseptics** These are poisons. They are used to kill bacteria. The early antiseptics such as carbolic acid (Chapter 9) were highly poisonous both to bacteria *and to humans*. They had to be used with great care. Most modern antiseptics are highly poisonous to bacteria and only slightly poisonous to humans, but we must take care to use them properly.

2 **Antibiotics** These are poisons that have been produced naturally by a fungus or similar

fig 19.2 *Multodisk being used to test the sensitivity of a bacterium (*E. coli*) to various antibiotics. The clear zones all have approximately the same diameter – what does this indicate?*

organism. They are not poisonous to the fungus that produces them, but are very poisonous to some other organisms, such as bacteria. To obtain antibiotics, the fungus is grown in tanks containing a solution of its food materials. The fungus makes the antibiotic, which passes into the solution. Then the antibiotic is extracted from the solution and purified. One useful property of antibiotics is that they are entirely harmless to most people yet are very poisonous to certain kinds of disease bacteria.

3 **Vaccines** When your body is invaded by disease bacteria or viruses, the lymphocytes of your blood (Figure 9.4) produce special substances, called **antibodies**. These act against the bacteria or viruses, or against the poisons they produce. The bacteria or viruses are unable to reproduce and infect the body, and the poisons are made harmless. When you *first* become infected by a particular kind of organism, you have no antibodies against the organism or its poisons: you become unwell and show the

fig 19.1 *Factory for the large-scale production of penicillin. On the right is the cover of one of the huge containers in which the* Penicillium *is grown.*

Alexander Fleming

stopped bacteria from growing there. Dr Fleming thought that the substance might be useful for disinfecting wounds. Tests made during the years following showed that the substance could kill many kinds of disease-causing bacteria. It was also found that the substance was not harmful to humans or other animals. Though the substance – now called **penicillin** – was known to have very useful properties, Fleming and his colleagues were unable to make it in large quantities; they could not make enough of it to treat even one person. Little was done after that until the time of the Second World War, when a group of scientists, led by Howard Florèy. Professor of Pathology at Oxford University, tried to develop methods for the large-scale production of penicillin. Working in Oxford and in the USA, they invented methods which eventually made possible the industrial production of penicillin. The discovery of this substance, and of ways of making large amounts of it, have saved millions of lives. In 1945, Sir Alexander Fleming, Sir Howard Florey and Dr Ernst Chain shared the Nobel Prize for Medicine as a reward for their work on penicillin. Sir Alexander Fleming died in 1955.

Following the success of penicillin, many fungi and other micro-organisms have been examined to see if they too produce **antibiotic substances** – substances that are active against disease-causing bacteria. One of the leaders of the search was Dr Selman WAKSMAN, of Rutgers University, USA. He and his team discovered many useful antibiotics, including streptomycin (used in treating tuberculosis) and neomycin. Dozens more antibiotics, each with its own specially useful properties, have since been discovered. They are an essential part of our defence against disease.

The story of antibiotics

Alexander FLEMING was born in Scotland in 1881. While working as a bacteriologist at St Mary's Hospital, London, in 1928, he noticed an unusual effect caused by a colony of the fungus, *Penicillium*. Certain kinds of bacteria, growing on the same dish of medium as the fungus, would not grow close to the fungal colony. It appeared that the fungus was producing a substance that went into the medium around the colony and

symptoms of the disease. A few days later, your lymphocytes have had time in which to make antibodies. Then the disease organisms are killed and their poisons are made harmless. You recover from the disease. After you have recovered, you still retain the ability to make antibodies against that kind of disease organism. The next time that the same type of organism tries to infect your body you can

immediately produce antibodies against it, and its attack on your body is unsuccessful. If you have had the disease once, and have recovered, you cannot be infected again. In other words, you are *immune* to that disease.

That is the *natural* way of becoming immune, but it has dangers, for the most serious diseases may damage your body permanently or may kill you before your lymphocytes have had time to make antibodies. It is safer to be made immune *artificially* by being treated with a vaccine. Then you do not risk the serious effects or the unpleasantness of the disease. A vaccine consists of *killed* bacteria or viruses, or *live* bacteria or viruses that are of a specially weak type that is unable to cause disease (see the section on vaccines against poliomyelitis). Another type of vaccine consists of bacterial poisons that have been treated chemically to make them non-poisonous. When you are treated with a vaccine (usually by injection), you do not become diseased, but your lymphocytes produce antibodies, just as if you really were infected. In this way you gain immunity without actually having the disease. If you have been immunised and the disease organisms enter your body, they cannot infect it, for you are immediately able to make antibodies against them and against their poisons.

Edward Jenner

Edward JENNER, an English physician, lived from 1749 to 1823, when the disfiguring and often fatal disease of smallpox was common. Jenner was a country doctor and noticed that people who had suffered from a similar, but mild, disease, called cowpox, did not catch smallpox. People normally caught cowpox from diseased cows when milking them. Jenner thought that if he deliberately infected people with cowpox he might give them protection against the dreaded disease of smallpox. In 1796 he began vaccinating people in this way, with very great success. His method, with small changes, has been used since then until the present day. Now, as a result of world-wide vaccination programmes, smallpox has been completely eliminated from almost every country in the world.

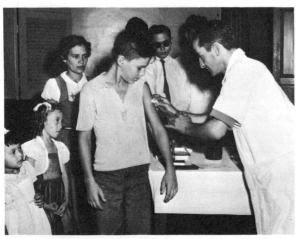

Vaccines are often used to protect the individual from certain diseases. They also protect the community for, if many people are immune, then epidemics cannot occur.

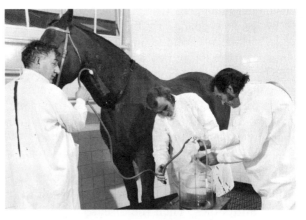

This horse has been injected with diluted poisons (toxins) made by diphtheria bacteria. In its blood, the horse produces antibodies (antitoxins) against the poisons. Some of the horse's blood is taken and made into a serum for injecting into a person who is dangerously ill with diphtheria. The horse's antitoxins help destroy the diphtheria toxins in the person's body.

Albert Sabin

Jonas Salk

The vaccines against poliomyelitis

Poliomyelitis is an unpleasant and crippling disease. It is easily spread from person to person: large epidemics can occur. One successful way of preventing the spread of the disease is by vaccinating young children. Once a person has been vaccinated, he is immune to the disease for life. The first vaccine was produced in 1953 by an American, Jonas SALK. The Salk vaccine contains poliomyelitis virus that has been killed by treatment with chemicals. People who have been injected with this vaccine become immune to the disease of poliomyelitis. Another American, Albert SABIN, developed a different vaccine for immunisation against poliomyelitis. The Sabin vaccine contains *living* virus which has been specially prepared to make it weak, so that it will not cause the disease. This vaccine is taken by swallowing –

usually a drop of the vaccine is placed on a sugar lump, which is then eaten. Persons taking this vaccine become immune. This living vaccine can spread from a vaccinated person to an *un*vaccinated person, so that the *un*vaccinated person also becomes immune to the disease. In this way a large proportion of the population of a country becomes immune, and epidemics are prevented from occurring.

The use of these two vaccines has led to a dramatic reduction in the number of cases of poliomyelitis in Britain, the USA and those other parts of the world where the vaccine has been widely used. Cases are few, and those few are usually not serious. But, now that the danger *seems* to be over, too many parents are not arranging for their children to be vaccinated. The number of unvaccinated children is increasing – and so is the danger of epidemics.

Things to do

1 Find out which antiseptics (or disinfectants) are used in your home or in your school, and for what purposes they are used. If possible, try some practical tests to find out how effective these antiseptics are.

2 Make up a set of rules for preventing the spread of disease organisms in

(a) the home
(b) the school
(c) public places such as cinemas, swimming baths, and buses.

3 Make a list of the diseases against which the members of your class are immunised. Draw a chart to show how many are immunised against each disease.

20 Plants – their food and our food

Green plants make their organic food in their leaves (Chapter 2). They transport it through the phloem to other parts of the plant. There it is used for:
(a) growth of new roots, stems, leaves, buds or flowers
(b) providing energy for all the activities of the cells of the plant
(c) building up a food reserve (or food store) for use in the future.

fig 20.1 *The food store of this beech seedling is in its thick fleshy seed-leaves (cotyledons). The first true leaves can be seen beginning to grow out from between the seed-leaves. In this plant, the seed-leaves have a small amount of chlorophyll and are able to photosynthesise. Later, when the true leaves are developed and the food store has been completely used, the seed-leaves wither away.*

Most plants store their food in seeds (Chapter 5). When the seed germinates, the seedling has a ready supply of food to provide materials and energy for its growth. Later, when it has grown roots and leaves, it is able to make its own supply of organic food. It becomes self-supporting.

Some plants build up large reserves of organic food in underground storage parts. These plants include:

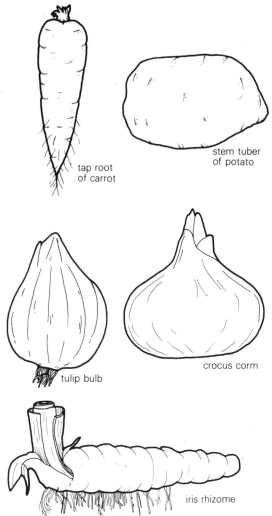

fig 20.2

During the growing season, increasing amounts of food are stored and the storage parts become large and swollen. Usually the store remains until the following growing season. Then the plant has a ready supply of food with which to make new stems and leaves. By having a food store, it gains the advantage of an *early start* at the very beginning of the growing season. It can spread its new leaves within a few days and begin its new season of growth much earlier than can those plants which have no food store.

fig 20.3 The flowers and leaves of snowdrops appear very early in spring. They grow from the corms which are below ground. At this stage, the trees have not produced their leaves and the other woodland plants are still leafless. The corm grows to full size and stores away its food supply for the next year before the trees make their leaves and cut off the supply of light to the snowdrop plants below. By making an early start in spring, the snowdrop is able to live in an environment that is shaded for the whole of the summer.

fig 20.5 The beans from the cocoa tree are used for making cocoa and chocolate. The beans contain large amounts of stored fat (cocoa butter); this is one reason why chocolate is a food with high energy content.

fig 20.4 Many animals (including ourselves) find that the food stored by plants is good for them too. (left) The carrot fly larvae feed on the growing plant. A treated plant is shown on the right. (above) How many different kinds of plant food store can you see? Can you think of any plant food stores that are sold in your local market?

Oil palm

Coconut palm

Soya plant

Sunflower plant
Maize plant

fig 20.6   Many plants store food in the form of oil. The oils
extracted from these plants have many uses: as cooking oil, as
salad oil and for making margarine. Some plant oils are used for
making soap.

Things to do

1   Find out what kinds of organic food material are stored in seeds and in the storage parts of plants. Test for starch, sugars, proteins and fats or oils. Your teacher will demonstrate how to perform the tests for these substances.

2   Follow the stages of growth of garden plants that have storage parts such as bulbs, corms, rhizomes, tap roots and tubers. Some can be grown in pots of soil in the laboratory; others can be grown in a garden bed. Note what happens to the storage part as the plant grows new stems, leaves and flowers. Note when and where the plant builds up its new food store, ready for the next growing season.

3   The food stores of plants are the main source of food for humans in all parts of the world. But in different parts of the world we feed on different kinds of plant. Find out which plant food stores are used as food for humans in different parts of the world. Here are some names of food plants to help you begin this project: wheat, rice, maize, millet, cassava, sago, potato, rye, barley, sweet potato, oats, yam. For each plant find out *where* in the world it is commonly eaten, *what part* of the plant the food is stored in and *what kind* of food material is stored there. Collect pictures and drawings of each kind of food plant – collect samples of the plant, or food made from it, if you can. You could put all the information, pictures and materials you have collected onto a large outline map of the world, making a display for your classroom.

4   We grow crops in our farms and gardens so that we can use their food stores to feed ourselves. Other animals, too, try to feed on the crops we have grown. Find out the names of insect and other pests of farm and garden crops. Collect examples of plants that have been attacked by these pests and display them in the laboratory.

21 Buds

A bud contains a group of cells able to grow rapidly when conditions are right. We call this group of cells a **meristem**. The meristem is closely surrounded by **bud scales**. The scales are often small; in some buds (for example, the 'eyes' of potato) they are almost non-existent. In other buds (for example, the Brussels sprout) they are relatively large and leaf-like.

The buds on the plants in Figure 21.1 have been inactive, or **dormant**, for the whole winter. They were made last year but only now, in spring, are they beginning to grow. Now they will produce new stems with leaves and flowers. The buds (or 'eyes') of the potato tuber produce new roots, too.

During the cold season the bud scales protect the meristem of the bud from cold and from the danger of drying out. The buds of such plants as carrot, tulip, daffodil, crocus, potato and many others are protected from cold because they are below ground. It is rare

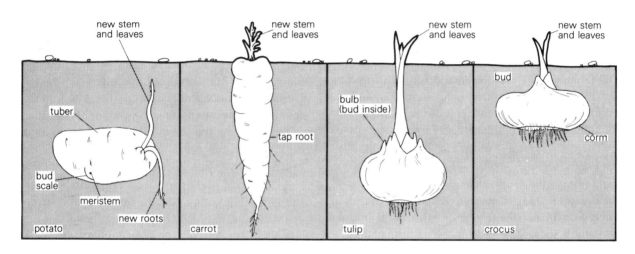

fig 21.1

fig 21.2   *A thin section through a bud of buttercup, cut lengthways and seen through a microscope.*

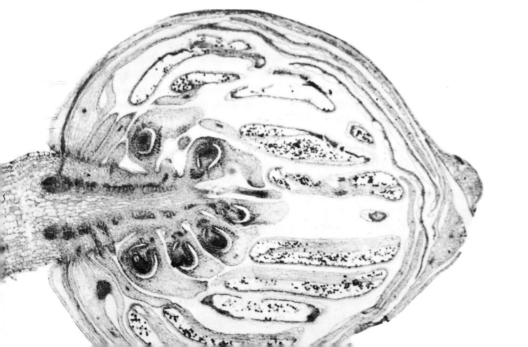

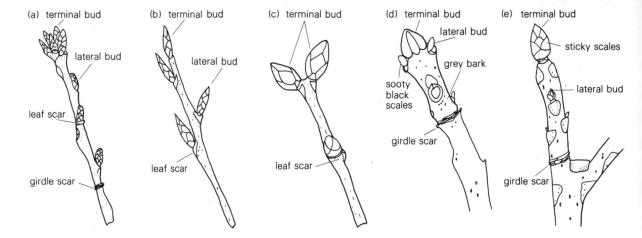

(a) terminal bud
lateral bud
leaf scar
girdle scar

(b) terminal bud
lateral bud
leaf scar

(c) terminal bud
leaf scar

(d) terminal bud
lateral bud
grey bark
sooty black scales
girdle scar

(e) terminal bud
sticky scales
lateral bud
girdle scar

fig 21.3

(except in the Arctic and Antarctic regions) for frost to penetrate more than a few centimetres below the soil surface.

Trees have buds. Can you say which kinds of trees have buds like those above?
A twig has one bud at its end (the **terminal bud**) and usually several buds along its side (the **lateral buds**). Usually the bud scales are thick and corky, to give protection to the meristem inside. Just below each lateral bud you may see a small scar on the bark of the twig. This shows where a leaf became detached from the twig and fell off, in the previous autumn. Every lateral bud has a leaf scar just below it, for the general rule is that buds are formed in *the angle between a leaf-stalk and the main twig*. We call this the **axil** of the leaf.

In the growing season the terminal bud opens. As the meristem grows to make a new length of twig, with leaves and flowers, the scales of the terminal bud fall off. They leave a ring of scars around the twig – the **bud scale scars** or **girdle scar**. If you look for the girdle scars, you can see the position of the terminal bud in previous years. You can find out how much the twig has grown each year.

The lateral buds produce new branches on the twig. They may remain dormant for several years before they do this.

Except for carrot, the plants mentioned in this chapter and in Chapter 20 live for many years. We call them **perennial** plants. By making buds, perennial plants protect their meristems during the unfavourable season. This helps them survive until the next growing season. The evergreen perennial plants (for example,

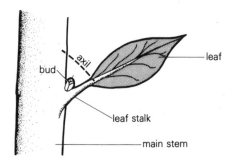

fig 21.4   A lateral bud is situated in the axil of a leaf

leaf
bud
axil
leaf stalk
main stem

fig 21.5   A thin section of a piece of twig of maple, cut length-ways and seen through a microscope. This section was cut in late summer, just before the leaf was due to fall from the twig. The abscission layer is a layer of cork forming across the leaf stalk so that it seals the wound caused by the leaf falling off. The lateral bud is in the axil of the leaf.

fig 21.6 In ash, the lateral buds are the first to open in spring; they produce very short stems crowded with tufts of simple flowers that have no petals.

fig 21.7 Rose-growers encourage strong growth by pruning back the stems; this leaves only a few buds to share the available food supply. Here a stem has been pruned just above a lateral bud which will later grow out to make a new branch.

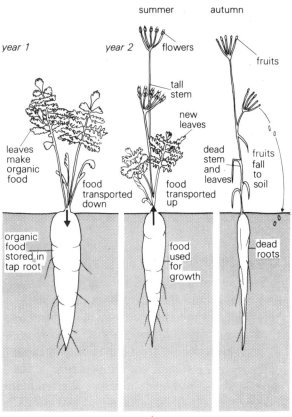

fig 21.8 The carrot lives for two years; in the first year it stores food in its tap root. At the end of the year we generally pull up the tap root and eat it. If the tap root is left in the ground, its food is used the following year to make stems, leaves and flowers. After flowering and making seeds, the plant dies. Plants like the carrot which go through their life story in two seasons, then die, are called **biennial** plants.

holly, laurel, yew, ivy, pine) always have leaves with which to make organic food to supply to the growing meristem. Other perennials, whose leaves fall off or die back at the end of the growing season, supply their meristems with food taken from the food store (Chapter 20) laid down during the previous year.

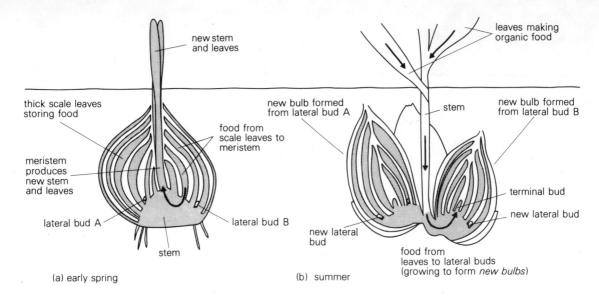

new stem
and leaves

leaves making
organic food

thick scale leaves
storing food

food from
scale leaves to
meristem

new bulb formed
from lateral bud A

stem

new bulb formed
from lateral bud B

meristem
produces
new stem
and leaves

terminal bud

new lateral bud

lateral bud A

lateral bud B

new lateral
bud

stem

food from
leaves to lateral buds
(growing to form *new bulbs*)

(a) early spring

(b) summer

fig 21.9   In early spring, the food stored in the fleshy scales of the tulip bulb is used for the growth of the terminal bud. In summer, the leaves make surplus food which is transported down the stem and to the lateral buds. Because each bulb has several lateral buds, *one bulb will have produced several new bulbs by the end of summer. In this way, the tulip not only survives from one season to the next, it also increases its numbers—it reproduces.*

Things to do

1   Look at a Brussels sprout (or cabbage) and a tulip bulb that have been cut in half lengthways. Find the meristem. Look for very small lateral buds in the axils of the bud scales.

2   During a cold frosty spell in winter or early spring, measure the temperature of the air just above soil level; measure the temperature of the soil at its surface and at various depths (2 cm, 4 cm, 6 cm and so on) below its surface. Repeat this in the same places on a warm sunny day later in spring. Compare the two sets of temperatures.

3   Follow the growth of the terminal bud of a twig from the time when it begins to open in spring or early summer. Make a chart to show the stages it passes through.

fig 21.10   Bryophyllum has an unusual way of reproducing itself. Buds grow at the edges of its leaves. Then the buds drop to the soil, where they take root.

22 The life of an insect

The life of an insect follows one of two patterns. Most of the insects we commonly find have the same life-pattern as the housefly and mosquito, as shown in Figures 17.5 and 17.4. The main features of their life-pattern are listed in this diagram (letters in brackets refer to the pictures below):

fig 22.1

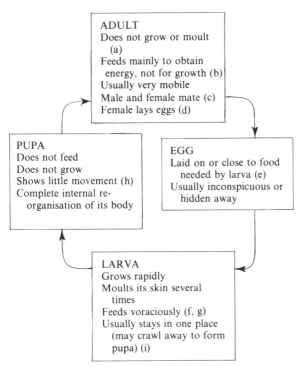

ADULT
Does not grow or moult (a)
Feeds mainly to obtain energy, not for growth (b)
Usually very mobile
Male and female mate (c)
Female lays eggs (d)

PUPA
Does not feed
Does not grow
Shows little movement (h)
Complete internal re-organisation of its body

EGG
Laid on or close to food needed by larva (e)
Usually inconspicuous or hidden away

LARVA
Grows rapidly
Moults its skin several times
Feeds voraciously (f, g)
Usually stays in one place (may crawl away to form pupa) (i)

(a) Like other adult insects, adult ladybird beetles do not grow. If you see ladybird beetles of different sizes, these belong to different species of the ladybird group of beetles.

(b) (above)   Adult butterflies, such as this painted lady butterfly, feed on the sweet, sugary liquid (nectar) found in flowers

(c) (below)   Adult male and female damsel flies mating on a reed by the water's edge

(d) *Female privet hawk moth laying eggs on privet*

(f) (above) *Larva of the great diving beetle* Dytiscus marginalis *feeding on a tadpole*

(g) (below) *Larvae of butterflies and moths have biting jaws and feed on leaves. This shows the head of a privet hawk moth larva, feeding on a privet leaf.*

(e) *Hard marble galls on oak are made by the tree when the cyndid wasp lays her eggs in the stem. The wasp larva feeds on soft tissue in the gall and pupates there. When adult, it bores its way out, leaving a hole, as seen in the photograph.*

In the other pattern of insect life story there is no larva or pupa. When the young insect hatches from the egg it looks like an adult, except that:

(a) it is smaller

(b) it has no wings

(c) it cannot reproduce

(d) the markings and colours on its body may not be like those of the adult.

We call this young insect a **nymph**. The nymph feeds and grows. As it grows, it needs

(h) Blowfly pupae are more rounded and darker in colour than the larvae they developed from, but they do not move or feed. Soon the hardened skins will be broken open and the adult blowflies will come out.

(i) Adult mosquitoes fly in the air and feed on blood (females) or plant juices (males), but the larvae live in water and feed on microscopic algae

wing bud   1 cm

to moult its skin several times, so that its body can increase in size. At each moult, it becomes more and more like the adult.

fig 22.2   These drawings show the nymph stage (or instars) of the grasshopper. In the later stages the nymph has wing buds. In the final stage, the adult has fully developed wings and can fly. It also has the ability to reproduce.

fig 22.3   (above) A dragonfly nymph (right) an adult dragonfly. How many differences can you see between these two stages in the life of the insect? The dragonfly is a carnivore: the nymph feeds on small aquatic animals and the adult feeds on insects which it often catches in mid-air.

23 Trees

The oldest living organisms are trees. In Britain, the yew grows for one thousand years or more. The oldest trees in the world are the bristle-cone pines in California. Some of these are over four-thousand-six-hundred years old.

Trees are perennial plants (Chapter 21). Unlike the smaller perennial plants (such as daffodil, dandelion or potato) trees do not lose the whole of their above-ground parts each autumn or winter. Their trunks, branches and twigs do not die back. These are more or less permanent; they are strong; they can resist the cold of winter. Even if a few of the smaller younger twigs do die back and fall off the tree, most twigs remain. These carry buds (Chapter 21) from which the tree can make more twigs, leaves and flowers when the next growing season comes. Each year the tree adds a little more to itself. It adds a little more each year for hundreds of years. After having grown for such a long time, it reaches a very great size.

The tallest living organisms are trees. In California, USA, the redwood trees grow up to one hundred metres tall.

fig 23.1   Redwood tree, California

fig 23.2   Yew belongs to the group of coniferous trees, but has no proper female cone. The single ovule swells to form a rounded brown seed in autumn, surrounded by a soft fleshy cup which is bright red in colour and attractive to birds. The leaves are relatively thick and covered with waxy cuticle, like those of most evergreen trees of northerly regions.

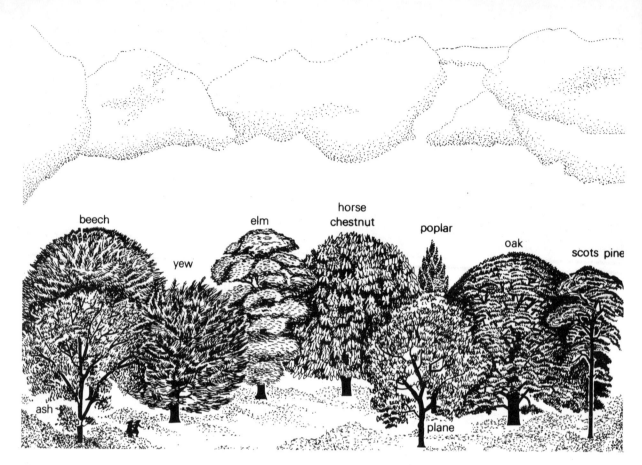

fig 23.3   How many of these trees can you find in your district?

The leaves of trees are *not* permanent, for they are fairly easily damaged by wind or frost, or they may be partly eaten by animals. Most trees make new leaves once every year – usually in spring or early summer. Some trees make leaves that are tough enough to last for several years before they finally fall from the tree. Holly, pine, cedar and yew are examples of trees with leaves that last for several years. Perhaps you can think of some more examples. If you look at a branch of one of these trees, you will see that the newest leaves are the cleanest and may be very shiny. If they are very new and are still growing to full size, they are smaller than the older leaves and may be lighter green in colour. The older leaves are generally dirtier especially if the tree is in a town

area. The very oldest leaves will probably be torn and spotted. They may be attacked by fungal disease, and there may be holes in them where they have been eaten by animals. These old leaves have served their useful life and will soon fall from the tree. But each year the new buds produce a new batch of leaves, so the old leaves are replaced by new ones. At any one time the branches of these trees carry new leaves, leaves that are one or two years old, and some older leaves – at all times the branches have leaves on them, even in the winter. Such trees are called **evergreen trees**.

Some trees lose *all* their leaves every autumn. In the winter, their branches are leafless. Oak, sycamore and apple are examples of trees of this kind. We call these **deciduous trees**. Can you think of some more examples of deciduous trees? The leaves of evergreen trees are usually rather thick and frequently covered by a thick

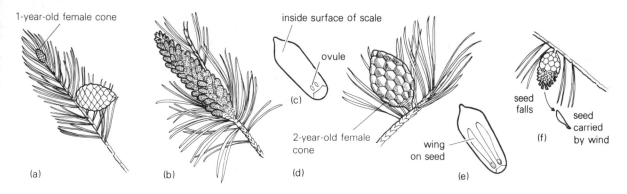

fig 23.4 (a) Young female cones of Scots pine (b) male cone of pine (c) scale taken from young female cone (d) older pine cone (e) scale taken from older cone (f) when the seeds are fully developed, the cone turns downwards and its scales curl apart.

fig 23.5 Larch trees in winter

shiny layer of a waxy substance, the **cuticle**. They are able to withstand the hardships of winter and attacks by animals for a few years, at least. But the leaves of deciduous trees are usually thin and delicate. They are made to last only for a few months, and the tree uses only the minimum amount of material and energy needed to make them.

Most trees belong to one of two main groups of seed-producing plants, the gymnosperms and the angiosperms. Gymnosperms have cones. Cones are either female or male. If you look at a young female cone and pull the scales apart, you will find two small **ovules** on each scale. The ovules are exposed to the air, so pollen grains can be carried in the wind from the male cones where they are made to land on the ovules in the female cones. Later, a nucleus (Chapter 3) from a pollen grain joins with an egg-nucleus in the ovule. Each ovule is fertilised in this way and develops to become a **seed**. After this has happened, the scales curl back, allowing the seeds to drop out of the cone. Trees which have cones are called **coniferous trees**, or **conifers**. Most conifers are evergreen, but some are not. For example, larch and swamp cypress are deciduous conifers.

In the angiosperms (or Flowering Plants, as

fig 23.6 Many trees have large colourful flowers: here we see flowers of horse chestnut in spring. Cherry, almond and apple also produce bright flowers and are often planted in gardens and parks for ornament. Contrast these with the small dull flowers of ash (Figure 21.6).

fig 23.7 In holly, the female flowers and male flowers are on separate trees. Only a tree with female flowers can produce holly berries in the autumn. The stamens of the male flowers produce the pollen needed to fertilise the ovules of the female flowers. Male trees produce no berries themselves, but we must always plant some male trees close to female trees if we want them to produce berries.

they are often called) the ovules are not exposed to the air, but are completely enclosed in an ovary. One or more ovaries, usually surrounded by stamens, petals and sepals, make up the flower. Pollen is usually transferred to the flower by insects or by wind from the stamens of another flower. It is held on the stigma, then grows a pollen tube through the tissues of the ovary. A nucleus from the pollen grain travels along the tube and joins with an egg-nucleus in the ovule. Then:

the fertilised *ovule* develops to become a *seed*

the *ovary* around it develops to become a *fruit*

Many of our common trees produce fruits. In autumn you can find the spinners of sycamore, the keys of ash and the acorns of oak. These are all **fruits**, each containing one seed. Conkers are the **seeds** of horse chestnut – on the tree they are contained in the thick-walled, prickly fruit. How many seeds does each fruit contain? The fruits from some trees are good to eat – can you think of some examples?

Things to do

1   Make a set of charts for all the common trees growing in your district. On each chart draw a sketch of the tree to show its shape; make a bark-rubbing and stick this on the chart; add some pressed leaves, a piece of twig and drawings (or specimens) of flowers and fruits. Write short notes about any ways in which we use the tree or its fruits and add these notes to the chart.

2   Try to grow some trees from seeds. You can gather fruits and seeds in autumn, and also obtain seeds from fresh fruits, such as lemon, orange or peach. Germinate the seeds in pots of soil in a greenhouse (if they are tropical) or outdoors. Later, you may be able to plant some of your seedlings. Then you will have young trees to study, and they can also help to decorate your school.

3   What foods do we get from trees? Make a

list. Then try to collect photographs of the trees and the foods we get from them. Try also to get samples of the foods. When you are making your list, remember that there are many tropical trees that give us food.

4 Trees also provide food for animals. Choose one kind of tree that is common in your neighbourhood and find out how many different kinds of animal feed on it. Find out what they eat – leaves, bark, roots, flowers, pollen, nectar, fruits, seeds. You could also include those animals that feed on the animals that feed on the tree. While you are studying these animals, note also those which obtain shelter in, on or around the tree.

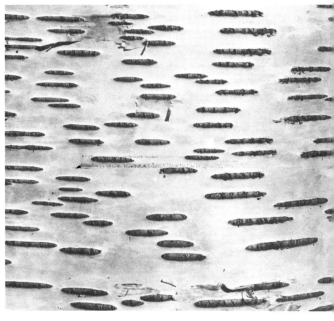

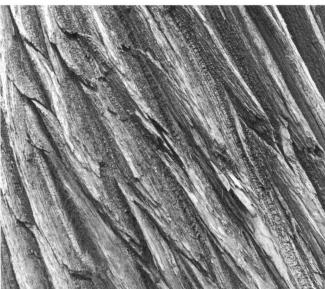

fig 23.8 Can you recognise these trees from their bark alone? The barks differ in colour – look for some trees of each type and find out what the colours are.

24 The plant eaters

In a wood there are trees, bushes and many different kinds of smaller plant. There is a layer of dead leaves and other dead parts of plants, which we call leaf litter. We may find moulds and larger fungi growing on the leaf litter. All these provide food for the plant eaters. Those animals that eat only plants are called **herbivores**. A wood can provide food for hundreds of different herbivores. The picture shows just a few common examples:

fig 24.2   Limpets graze on the thin film of algae growing on the surface of the rock

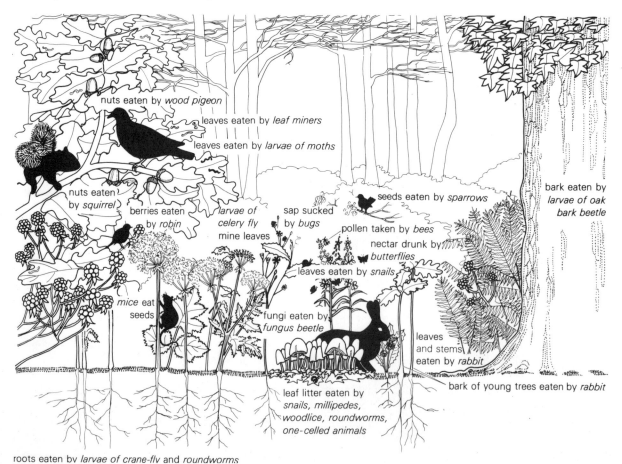

nuts eaten by *wood pigeon*

leaves eaten by *leaf miners*

leaves eaten by *larvae of moths*

nuts eaten by *squirrel*

berries eaten by *robin*

larvae of celery fly mine leaves

sap sucked by *bugs*

seeds eaten by *sparrows*

pollen taken by *bees*

nectar drunk by *butterflies*

leaves eaten by *snails*

bark eaten by *larvae of oak bark beetle*

mice eat seeds

fungi eaten by *fungus beetle*

leaves and stems eaten by *rabbit*

leaf litter eaten by *snails, millipedes, woodlice, roundworms, one-celled animals*

bark of young trees eaten by *rabbit*

roots eaten by *larvae of crane-fly* and *roundworms*

fig 24.1   A few of the many plant-eating animals that you can find in a wood

fig 24.3   Snails and slugs scrape at the surface of plant material, using their flexible, toothed 'tongue' which you can see magnified here. In the garden, slugs cause serious damage to the plants they eat. Pond snails feed on water plants and also scrape away the thin film of algae growing on the surface of rocks (or on the glass of an aquarium).

fig 24.4   Many tropical trees are tall. Their fruits can be reached only by animals able to fly (birds, insects) or climb. This Amazon squirrel monkey spends its time in the tree-tops, finding all its food there.

fig 24.5   A plague of locusts arrives. In a short time, the plants in the area will be stripped of all their leaves and the crops ruined.

Things to do

1   Visit a wood, looking for herbivores. Make a chart like the drawing opposite, but add pictures and the names of the herbivores you have found.

2   Make similar charts for other places where herbivores feed: a park or garden, a hedge, a pond, an aquarium, or any other place you know of that it is convenient for you to visit and study.

3   Some of the animals in the drawing eat more than one kind of food. Some of them feed partly on plants and partly on animals. These we call **omnivores**. Which of the animals shown in the drawing are omnivores? Name some omnivores you have found in other places (see 2 above).

25 Timber!

How many things in your classroom are made from wood? Trees provide us with one of our most widely-used building materials. We use timber for making houses, sheds, furniture, bridges, boats and even aeroplanes. For many purposes timber is superior to steel, for it is cheap, light and strong. Many timbers can be bent without snapping, for the wood is springy. Other timbers can be finished to show the beautiful patterns and colours of the grain; with such timbers our houses can be made more pleasant to live in.

The great strength of wood gives us a clue to the reason why trees can grow to such large sizes. Their branches resist damage by wind, for they are strong yet springy. They do not break if, in winter, they are weighed down by a thick covering of snow. Year by year, the tree adds more branches to itself, yet only a few are lost through damage by wind and weather. So the tree can gradually grow to great size. As the tree becomes larger, the weight of its branches and leaves gradually increases. The tree trunk must become thicker to support this increasing weight. When a tree has been cut down, we can see that most of the trunk consists of a solid column of wood. The wood consists mainly of xylem cells (Chapter 3) which have thick strong walls, but no living contents. These are what give the trunk its great strength. The outer xylem cells are gen-

fig 25.1 *Many items of sports equipment are made from wood. Which kinds of wood are used in making the equipment illustrated here? Different parts of the equipment may be made from different woods, each with its own special properties. Try to work out which* special properties — springiness, lightness, hardness, stiffness, cheapness, workability (ease of cutting to complex shapes), resistance to rotting — are needed for each item of equipment.

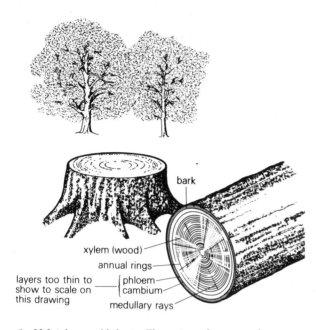

fig 25.2 (above and below)   The regions of a tree trunk

erally used for transporting water from the roots to the upper part of the tree. Among the xylem cells are strands of cells of a different kind. These strands are the **medullary rays**.

Around the wood is the growing region, the **cambium.** These are living, dividing cells. Each spring and summer they produce new cells, most of which are added to the outside of the woody column, increasing its diameter. In a cross section the newly-added xylem cells appear as a ring around the outside of the wood. The cells produced in the spring are usually larger than those produced later in the year. The effect of this is to make a ringed pattern, in which one ring represents one year's growth. These are called **annual rings**. The age of the tree can be found out by counting the number of annual rings. This is how the age of the bristlecone pines of California has been discovered (Chapter 23).

The outside of the trunk is covered with bark. This protects the tree from cold because it is a good heat-insulator. We make use of the heat-insulating properties of bark when we use floor-tiles and table-mats made from cork. Cork is the bark taken from a special kind of tree that grows in southern Europe, called the cork-oak. The cells from which bark is made have thick walls, waterproofed by a fatty substance. This makes bark waterproof, and we make use of this property when we use cork for making the stoppers of bottles.

As the trunk grows thicker, the bark around it becomes stretched and tends to split. The bark has its own cambium, which grows to make new bark to seal the splits as they occur. In this way, the tree is always covered with a continuous layer of bark. This is important because bark prevents the inside of the tree from becoming dry in hot or windy weather. It also helps to prevent disease organisms, such as fungi, from attacking the tree. If the bark is removed or damaged by cuts, disease may enter the wood and the tree may become infected and die. The natural splitting and growth of the bark produces the various patterns of

2 Discover what is being done in your area to conserve trees. Find out about schemes for planting trees in parks and gardens and along roads and motorways. Find out about reforestation schemes. Find out about nature reserves. Find out about local and national organisations for preventing large-scale destruction of our trees (including those in hedges). Are there any organisations you could join? Is there anything practical that you could do to help conserve our trees?

3 Make a display of products that are obtained from trees. Include timbers, foods, oils and waxes, gums and resins, perfumes, flavourings, dyes and tannins, medicines, rubbers and other useful materials.

fig 25.4 *Each year, millions of trees are felled to provide the wood-pulp needed for making paper. Although re-forestation is practised in many countries to provide further supplies of wood-pulp, the demand for paper rises faster than ever. One way of reducing this excessive use of the world's forests is to recover waste paper and card, re-pulp it and use the pulp to make new paper. Using recycled paper is one way of conserving our trees.*

fig 25.3 *After pruning branches from this rose tree, the gardener is painting the exposed wood with special sealing compound to keep out the spores of disease fungi*

markings on the bark, by which we can recognise many kinds of tree.

The life-style of a tree is to grow a little extra wood each year for tens, hundreds or even thousands of years. Trees grow larger little by little. What a pity to cut down a tree unless we really must! A few minute's action with an axe can destroy a plant that has lived for a hundred years or more. It will take a hundred more years to replace it.

Things to do

1 Ask your handicraft department to show you samples of different kinds of wood, and to explain their special uses.

fig 25.5 Trees provide food and shelter for wild-life. When the banks of motorways are planted with young trees, the landscape is improved and a nature reserve is also formed where people cannot interfere with the wild-life. Here, schoolchildren are taking part in a tree-planting scheme.

26 Eating

We have two sets of teeth. When we are young (six months to two years old) we grow our first set—the milk teeth.

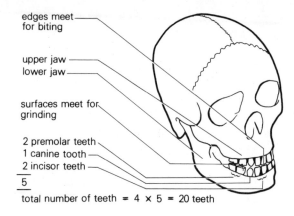

fig 26.2

edges meet for biting
upper jaw
lower jaw
surfaces meet for grinding
2 premolar teeth
1 canine tooth
2 incisor teeth
5
total number of teeth = 4 × 5 = 20 teeth

Our teeth are of four main kinds:

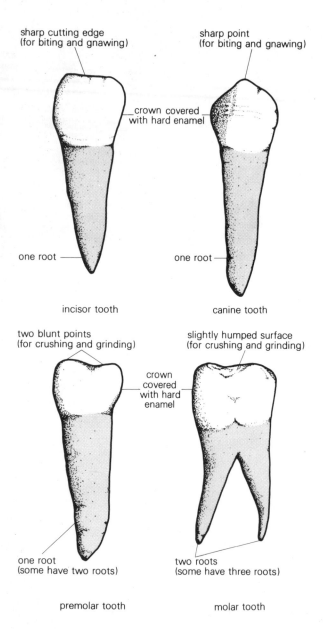

fig 26.1

sharp cutting edge (for biting and gnawing)

sharp point (for biting and gnawing)

crown covered with hard enamel

one root

one root

incisor tooth

canine tooth

two blunt points (for crushing and grinding)

slightly humped surface (for crushing and grinding)

crown covered with hard enamel

one root (some have two roots)

two roots (some have three roots)

premolar tooth

molar tooth

As we grow older, the milk teeth drop out one by one, being replaced gradually by teeth of the second set, the permanent teeth. This happens between the ages of about six years and seventeen years.

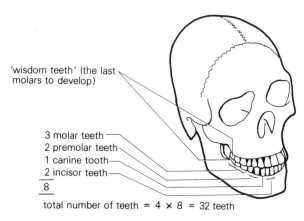

fig 26.3

'wisdom teeth' (the last molars to develop)
3 molar teeth
2 premolar teeth
1 canine tooth
2 incisor teeth
8
total number of teeth = 4 × 8 = 32 teeth

The lower jaw is *loosely* hinged to the skull. We can open and shut the jaw; we can move it slightly sideways and also backwards and forwards. The sideways and backward-forward motion allows us to chew food between the premolars and molars of opposite jaws.

Our teeth are suitable for eating plant matter of many kinds, especially soft ripe fruit. Possibly our ape-like ancestors fed on the fruits of the trees in which they spent much of their life. Perhaps they ate the softer leaves and any soft-bodied insects and small reptiles they could catch. Our teeth are *not* good for eating raw flesh. Raw flesh cannot be crushed or ground into small pieces, because its texture is too rubbery. Raw flesh must be cut into small pieces, but our teeth cannot do this, except by repeatedly biting off small pieces with our incisors. Fortunately, we have invented knives for cutting meat and have discovered how to make meat softer by cooking it.

Herbivorous mammals, such as cows, sheep, goats and horses, eat nothing but plant matter. This kind of diet raises some problems:

(a) Most plant matter contains a high percentage of water and much indigestible woody and fibrous material that has little or no nutritional value. Herbivores need to eat a *large quantity* of plant matter to obtain a given amount of nourishment. Many spend almost all their waking hours feeding.

fig 26.5 (right)   The gut of a rabbit. Plant matter takes a long time to digest, so the small intestine of herbivores is relatively longer than that of other kinds of mammal. In the caecum and appendix of rabbit, the cellulose from plant cells is digested by bacteria that have a cellulose-digesting enzyme. Some of the products of this digestion can be used by the rabbit. In return, the bacteria have food, shelter and a place in which to live.

fig 26.4 (below)   Molar tooth of an elephant. The hard enamel around the edge of the tooth is not easily worn away. Between the enamel, there is a softer, bony substance (dentine) which wears more easily. Because the enamel projects above the softer dentine, the tooth has a strongly ridged surface that is ideal for grinding.

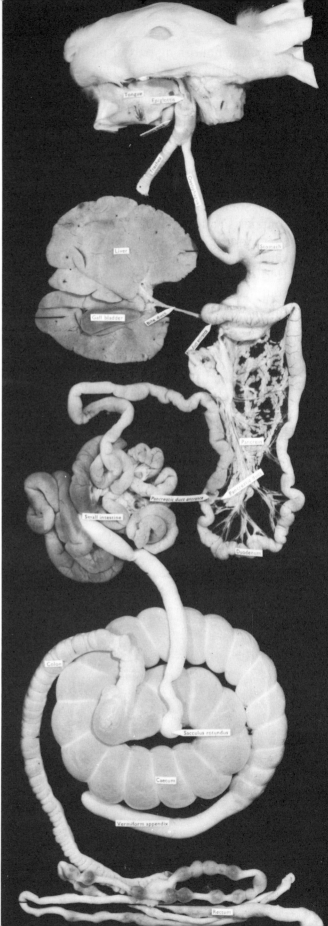

(b) The nourishing part of plant matter is the cytoplasm and cell sap. This is surrounded by the cell wall (Chapter 3), which is made of cellulose, which cannot be digested. To aid digestion, the cells of plant matter must be broken open, by **grinding** and **chewing**. Herbivores must have premolars and molars which are well developed for grinding and chewing plant matter.

(c) Some plant matter (for example, grass) contains small grains of silica. These hard, sharp grains, like microscopic grains of sand, cause much wear to the teeth of herbivores. Herbivorous mammals have teeth that grow continuously, to make up for the very high rate of wearing away.

Things to do

1 Look at skulls and teeth of herbivorous mammals such as cow, rabbit, sheep, guinea-pig, and horse. If possible, watch live animals feed, or look at a film of them feeding. Then answer these questions:

(a) Which of the animals that you have looked at have canine teeth?

(b) Which of the animals have no incisor teeth in their upper jaw?

(c) In which direction does (i) the rabbit, (ii) the cow, (iii) the guinea-pig and (iv) the horse, move its jaw while chewing?

(d) A rabbit's incisor teeth must be kept sharp, yet the food it eats could soon make them blunt. How are a rabbit's incisor teeth kept permanently sharp?

(e) The premolar and molar teeth of herbivorous mammals are set close together to form a compact group of grinding surfaces on each side of each jaw. The grinding surface is ridged (Figure 26.4). In which direction do the main ridges run in the animals named in question (c)? What can you say, in general, about the direction of chewing movements and the direction of the ridges?

(f) Describe the way in which a cow eats grass.

2 Make a collection of skulls and jaws of small mammals, such as mice, voles, bats, moles and shrews. These may be found among leaf litter in woods and under hedges; if you can find the pellets of owls or rooks, you will be able to identify the pieces of skull and jaw of small mammals that the bird has eaten. Find out what kinds of food these small mammals eat — many of them are not herbivores. Do their teeth show any adaptation to their diet?

27 The meat eaters

Tigers, lions, eagles, alligators, wolves, sharks — there are many large meat-eating animals in the world. We call them **carnivores** because their only food is the flesh of other animals. The world's largest carnivore is a lizard, the Komodo Dragon, from Indonesia. It reminds us of the great reptile carnivores of the past during the time when the dinosaurs lived. But not all dinosaurs were carnivores — and not all carnivores are large. A centipede is a carnivore, for it feeds only on other animals. There are carnivores even smaller than the centipede, for example *Didinium* (Chapter 4).

fig 27.1 (above)   *A Komodo dragon*

fig 27.2 (below)   *A devil's coach-horse*

The main feature of the carnivore's lifestyle is that it catches, kills and eats other animals. Some meat-eating animals, such as the raven, the crow and the devil's coach-horse (actually a beetle!), wait until they find an animal which is *already* dead, and then eat that. We call these **carrion-eaters**, and do not think of them as being true carnivores. The larva of the sexton beetle is another example of a carrion-feeder. Carrion-feeders will not be discussed again in this chapter. Here, we are concentrating on carnivores that hunt and kill, and then eat.

What does an animal need to be a successful carnivore? Here are the main features of the carnivore lifestyle:

1   **A carnivore must be able to find where its food is:**   This may not be easy. Except in desert areas, a herbivore usually knows where to find a patch of plants to graze on. The plants cannot run away to a different place tomorrow! But a carnivore, hunting for a rabbit, first has to find it. It may be very hard to see; it may be inside its burrow. Even if it is in the open, the colouring of its fur makes it difficult to detect by eye. Carnivores need very good senses. Some have good eyesight. Have you ever watched a kestrel hovering above a meadow? It is watching for mice or other small animals. From that great height, a human could probably not see them, but the bird can see them easily. After a short while, it spots a mouse and swoops down, to catch it, kill it and eat it. Dogs are carnivores, but some have rather poor eyesight. They rely on a very good sense of smell, instead. Pit-vipers, such as the rattle-snake, have a pit just below each eye. These pits contain sense-organs able to detect heat. The snakes hunt for food at night. By detecting the heat given out by small warm-blooded animals, such as mice and birds, they can easily locate their prey in darkness. You can find several different kinds of spider in your garden, and these are all carnivores. The zebra spider is hard to catch, for it jumps away very suddenly and quickly as you try to touch it. Its rapid jumping helps it to catch smaller

animals. It creeps after them and then suddenly jumps at them. Wolf spiders have a different way of catching food. They can run very fast, so are speedy enough to chase and catch their prey. Other kinds of spider do not run around trying to catch up with their prey. They stay in one place and let the food come to them. They build a web. When an insect or other small animal becomes caught in the web, the spider is waiting near by, ready to collect and eat anything caught in the web.

2 **A carnivore must be able to catch its food:** Like the wolf spider, most carnivores can run at high speed. Among carnivores we have the fastest-running animal in the world, the cheetah, which is capable of 100 km per hour. Two fast-flying birds, the swallow and swift, combine good eyesight and rapid accurate flight, to make them expert at catching flying insects in mid-air. Many dragonflies too are fast, well-controlled fliers and, often take their prey in mid-air.

Running or flying at high speed requires much energy. Many carnivores save energy by getting as close as possible to their prey before they are seen. They creep close, very quietly, and then suddenly rush at the animal from a short distance. Lions, tigers, pythons and many others act like this. They are helped by the colouring and marking of their coats. They are camouflaged to match their natural back-

ground, so they can come very close to their prey without being noticed. Frogs and toads are not suited to chasing other animals at high speed. They sit and wait until an insect or worm comes past. They are camouflaged and, if they stay absolutely still, they cannot be easily seen. When its prey is in reach, the frog flips out its long tongue. The tip of the tongue (but not the whole frog) *moves very quickly*. It is sticky. The prey is caught on the tip of the tongue and carried back into the frog's mouth. In a similar way, a pike may wait hidden in the shade of floating water weed, ready to pounce on any small fish that happens to pass close by.

3 **A carnivore must be able to stop its prey from escaping:** Many carnivores have long sharp teeth with which they can hold and kill their prey. The canine teeth of the dog are good examples. Owls, kestrels and eagles have strong bills, and their upper bill has a downwardly curved tip that is sharply pointed. The claws on their toes are long, curved and sharply pointed. Bill and claw are excellent weapons for attacking and killing prey. Octopus and squid have suckers on long tentacles; the tiny hydra found in ponds (Chapter 29) has tentacles covered with special cells that stick to or wind around an animal it has caught. Some of these cells inject a poisonous substance into the prey, to paralyse it. The captive is unable to

fig 27.3 *How a frog catches an insect. Frogs and older frog tadpoles are carnivorous, but the young frog tadpole is herbivorous.*

It scrapes at the thin film of algae growing on rocks or stones and on the stems and leaves of water plants.

fig 27.4 *The teeth of carnivorous fish and reptiles like this crocodile are of little use for cutting up the prey into small pieces. Their main use is for gripping the prey to stop it from escaping.*

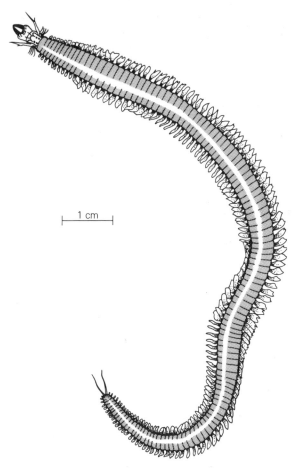

struggle to free itself and, before it can recover, it is swallowed whole. A python wraps itself tightly around its victim. The prey cannot get away and cannot breathe. After the prey has died through lack of air, the python swallows it whole.

4 **A carnivore must be able to digest a large meal:** There is no point in a large carnivore such as a lion using up lots of energy in chasing a mouse. To be worth chasing, the prey animal must be large enough to provide (when eaten and digested) at least as much energy as was spent in catching and eating it. After a python has swallowed its prey, the digestion of the dead animal usually takes several months. Digestion would be quicker if the prey could be cut into small pieces before being swallowed but special teeth are need for cutting

fig 27.6 *The ragworm lives in sand on the seashore. It feeds by putting out its proboscis, which has teeth on it. These grip the prey, then the proboscis is pulled back into the worm, with the prey attached.*

raw flesh, and the snakes do not have teeth of this kind. The best kind of meat-cutting teeth are the **carnassial teeth** of members of the dog and cat families (the latter includes lions). The carnassial teeth are molars or premolars, but they do not have the ridged, grinding *surface* that we find in herbivores (Chapter 26). They have a sharp, toothed **edge**. The edge of one carnassial tooth slides across the edge of the carnassial tooth of the opposite jaw. This gives a **scissoring** action. The carnassial teeth and other molar and premolar teeth all have points on them; these stop the flesh from sliding out of the scissors. The arrangement is just like a

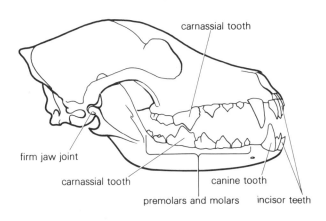

fig 27.5 *The skull and teeth of a dog*

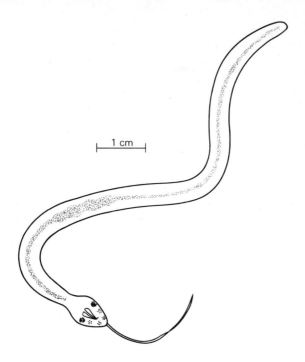

fig 27.7   Ribbon worms are common on the seashore. Their proboscis can be shot out quickly to wrap around the prey (which might be a ragworm). Some ribbon worms have a sharp needle on the proboscis to inject paralysing poisons into the prey. Other worms have a sticky substance on the proboscis to prevent the prey from escaping.

pair of kitchen scissors, with serrated blades. If the screw that holds the blades of a pair of scissors is loose, the scissors do not cut properly. This explains why the joint between the lower jaw and skull of a carnivorous mammal is very firm. The jaw cannot be moved from side to side or backward and forward, as it does in the chewing action of herbivores. So dogs, cats, lions and similar carnivores cannot

chew their food. This does not matter, for raw flesh is not chewable. Carnivores' jaws snap open or shut with a firm scissoring action suited to cutting raw flesh into smaller pieces before it is swallowed.

Things to do

1   Find out what you can about these carnivores:
(a) *Drosera* (Chapter 8), or any other plants that feed on insects
(b) pirana, a fierce tropical fish
(c) chamaeleon, a tropical lizard
(d) Portuguese man-of-war, often washed up on our beaches
(e) the sparrowhawk
(f) the weasel
(g) centipedes
(h) the otter
(i) the common shrew
(j) the kingfisher
(k) the cormorant
(l) the barn owl
(m) the red-backed shrike
Decide which special carnivore features these animals possess.

2   Collect drawings or photographs of all the different animals mentioned in this chapter. Make a display for the classroom.

3   Collect live specimens of carnivores from a local pond or stream. Set out a display in the laboratory. Label the display to point out the features of the animals that suit them to their lifestyle.

28 Roots

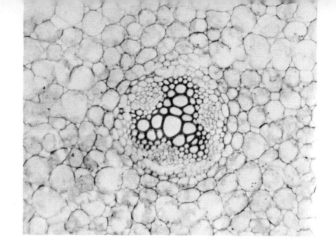

Roots have three main jobs to do. They:

1 **Anchor** the plant in the soil, so that its stem can grow upright and remain upright in all weathers.

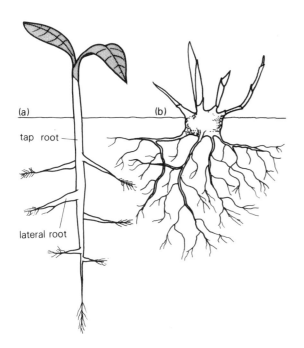

fig 28.1 *Two kinds of root system: (a) tap root system; there is one main tap root, with narrower lateral roots branching from it (b) fibrous root system; all roots have about the same diameter; several originate from the base of the stem and have lateral branches.*

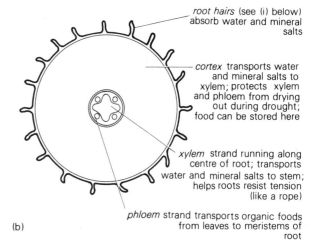

root hairs (see (i) below) absorb water and mineral salts

cortex transports water and mineral salts to xylem; protects xylem and phloem from drying out during drought; food can be stored here

xylem strand running along centre of root; transports water and mineral salts to stem; helps roots resist tension (like a rope)

phloem strand transports organic foods from leaves to meristems of root

(b)

2 **Absorb** water and soluble mineral salts from the soil, for use by the rest of the plant.

3 **Transport** water and soluble mineral salts to the upper parts of the plant.

The roots of some plants are used for food storage (for example, carrot, Chapter 20). Some plants have nodules on their roots, containing nitrogen-fixing bacteria (Chapter 8).

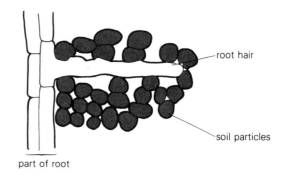

root hair

soil particles

part of root

(c)

fig 28.2 *(a) (Top) A cross section through a young root of a member of the buttercup family, seen through the microscope (b) explains what jobs are done by the different tissues of the root (c) section through part of the root showing the root hair*

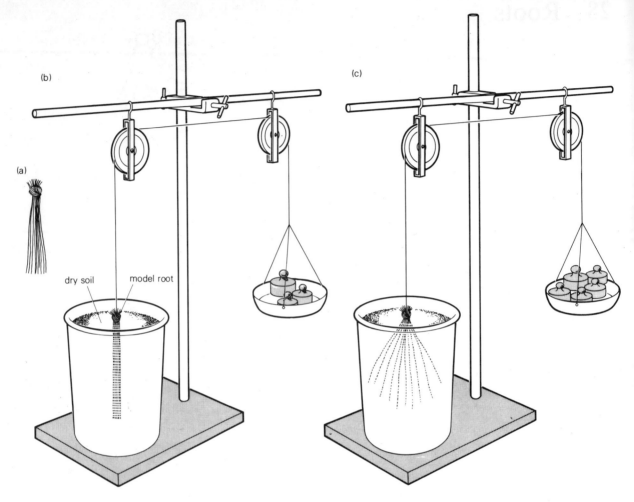

(b)

(a)

dry soil   model root

(c)

fig 28.3 See 'Things to do' 2: (a) make a model root from ten pieces of string, each 20 cm long, knotted together (b) bury the 'root' with its branches close together; add weights to the pan gradually *and find out how much weight is needed to pull the 'root' out of the soil (c) repeat after burying the 'root' with its branches well spread through the soil.*

Things to do

1 Germinate some mustard (or other) seeds on damp paper tissue. Keep the dish covered so that the air around the growing roots is damp. Examine the root hairs using a lens. On what region of the root are root hairs formed?

2 Using a model root made from string (Figure 28.3) investigate the anchoring power of a well-spread root system.

3 Sow some mustard (or other) seeds in loose soil. When they are about 5 cm high, hold them by their stems and *gently* pull the plants out of the soil. What does this show you about one of the jobs of the root hairs?

4 Roots can grow from a piece of stem if its lower end is kept in water or damp soil. Gardeners use this method for obtaining many new plants from cuttings taken from a few plants. Take cuttings from garden plants, including some shrubs. Woody cuttings usually root better if they are treated with a special rooting hormone. Follow the instructions on the packet or bottle.

29 Plankton

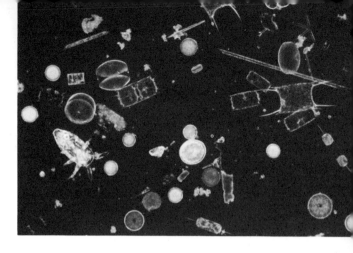

The seas, lakes, rivers and ponds of the world contain millions upon millions of small organisms. Most of these float in the water, usually at or near the surface. Some can swim but, since they are small, their powers of swimming are slight, and they are mostly swept along in the currents. These organisms make up the plankton. They include:

Simple plants, many one-celled, belonging to the group of Algae. These plant-plankton (or phytoplankton) are the primary producers (Chapter 4) of the waters of the world. The seas cover such a large area of the Earth's surface and are so rich in plant plankton that the total amount of photosynthesis of organic materials by plant plankton is greater than that of all the land-living plants of all kinds. Plant plankton can not live below a depth of about 100 m because the amount of light penetrating below 100 m is not enough for photosynthesis. In muddy water, they may be restricted to the upper 10 m.

Simple animals, many one-celled protozoans and large numbers of different kinds of shrimp-like animal. The animal plankton (or zooplankton) also includes the eggs and larvae of aquatic animals, small jellyfish and similar animals and young fish (including young herring, see below). Some of the animal plankton are herbivores, feeding on the plant plankton; some are carnivores, feeding on the smaller members of the animal plankton; others are omnivores, feeding on small plankton of all kinds.

Plankton forms the food of many of the larger water animals. Individual plankton organisms are very small – many kinds are almost invisible to the eye – but they make up for this by existing in extremely large numbers.

fig 29.1 *Tiny plankton form the only food of the largest animal species that has ever existed – the Blue Whale. An adult Blue Whale weighs almost 150 tonne (equal to the weight of twelve elephants). It is estimated that a Blue Whale eats 450 tonne of plankton each year – mainly krill, which are shrimp-like organisms, up to 3 cm long, and are the commonest and largest of the animal plankton. The whales take in mouthfuls of seawater and filter the krill from it. On their upper jaw, the whales have plates of whalebone which have fringed edges. The water passes out between the fringes and returns to the sea; the krill are retained in the mouth and then swallowed.*

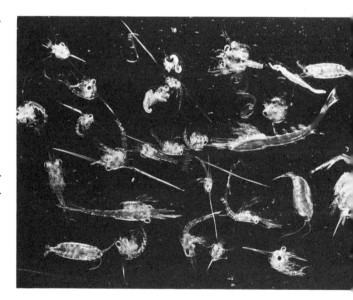

The animals that feed on plankton do not chase individual plankton organisms — they simply filter them from the water by the thousand.

The waters of the Arctic seas and North Atlantic are well stirred by storms. Plankton are plentiful and support a large population of fish. These seas are good fishing areas that are

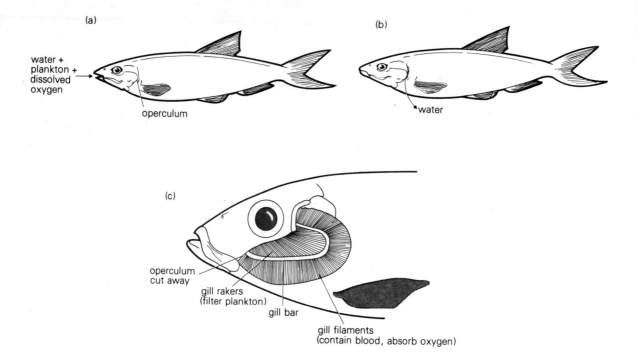

fig 29.2 *The herring combines getting food with getting oxygen. It takes in a mouthful of water and passes it out through its gills. Oxygen is extracted from water. At the same time, plankton are filtered from the water by the network of spiny gill-rakers. The plankton are then swallowed.*

exploited (and often over-exploited) by fishermen. Like the filter-feeders, we do not try to catch individual herring or cod – we just use nets to strain them from the water by the thousand (Figure 11.6).

fig 29.3 *Like the herring, the mussel uses its gills to filter its food from the water. The mussel is unable to move around quickly yet, a filter-feeder, it is able to obtain all the food it requires while remaining in one spot.*

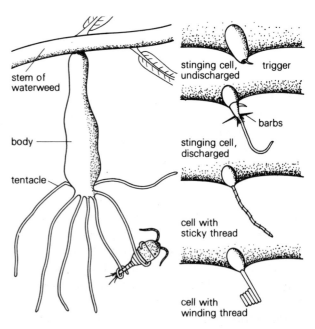

fig 29.4 Hydra *lives in fresh water, usually attached to the underside of the leaves of water plants. Its tentacles dangle in the water. Any small animal that swims by, or is carried past in the current, and accidentally makes contact with a tentacle, is immediately caught by the special cells on the tentacles.*

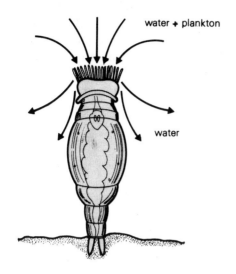

water + plankton

water

fig 29.5 (left)   *Rotifers are common in fresh water. Like the mussel,* Hydra *and many filter-feeders, they tend to remain in one place and collect their food from the water around them. Rotifers have tiny cilia at their 'head' end that beat rapidly, producing feeding currents in the water. Plankton are carried towards the rotifer in these currents and are then swallowed. The mosquito larva is another example of an animal that makes feeding currents.*

fig 29.6 (below)   *(a) In shallow water, the dead bodies of plankton fall to the bottom and are decomposed; the materials they contain are released into the water for use again (b) In deep water, most of the dead plankton fall far below the level in which the plant plankton live (c) Ocean currents, the action of wind and storms, bring the mineral material to the surface, where it is used by the plant plankton. The water is rich in plant plankton, which supports a rich animal plankton; many larger animals come to these waters for the good supply of plankton they find there. Many fishermen come to these waters for the plentiful supply of fish they find there.*

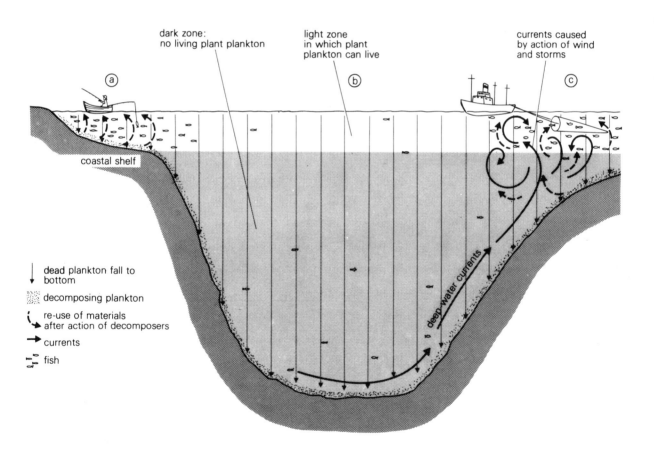

dark zone:
no living plant plankton

light zone
in which plant
plankton can live

currents caused
by action of wind
and storms

(a)

(b)

(c)

coastal shelf

deep-water currents

↓ dead plankton fall to bottom

decomposing plankton

re-use of materials after action of decomposers

→ currents

fish

30 Plants, animals and the atmosphere

A suitable atmosphere is one of the essential living conditions for all organisms (Chapter 1). For almost all organisms this means an atmosphere containing a relatively high percentage of oxygen. The Earth's atmosphere contains about 21 per cent oxygen, and this is enough

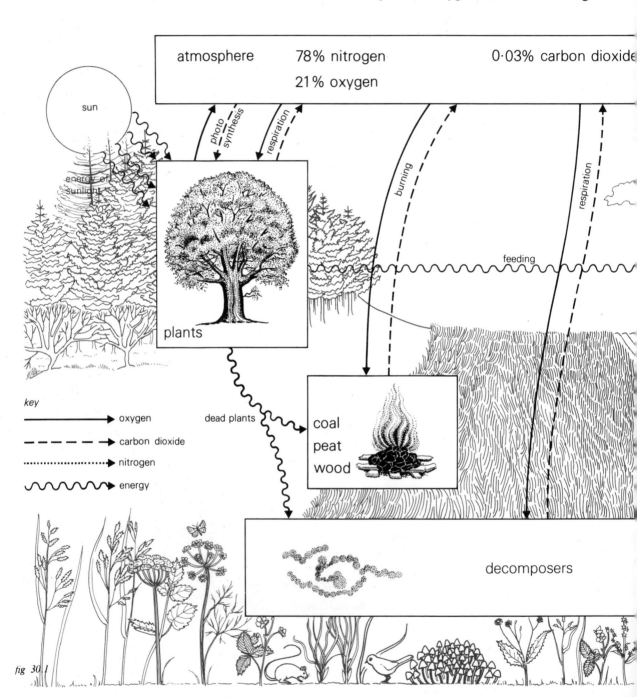

fig 30.1

to supply the oxygen that organisms need for respiration. When they respire, organisms release energy stored in chemical form, as organic materials such as starch and sugar. The energy is converted to other forms, such as movement of muscles or cilia, heat or light (can you think of examples?) or it is used for driving the many activities that take place inside cells, and for providing the energy needed to build up complex organic materials from simpler ones. Without a ready supply of oxygen from the atmosphere, most organisms

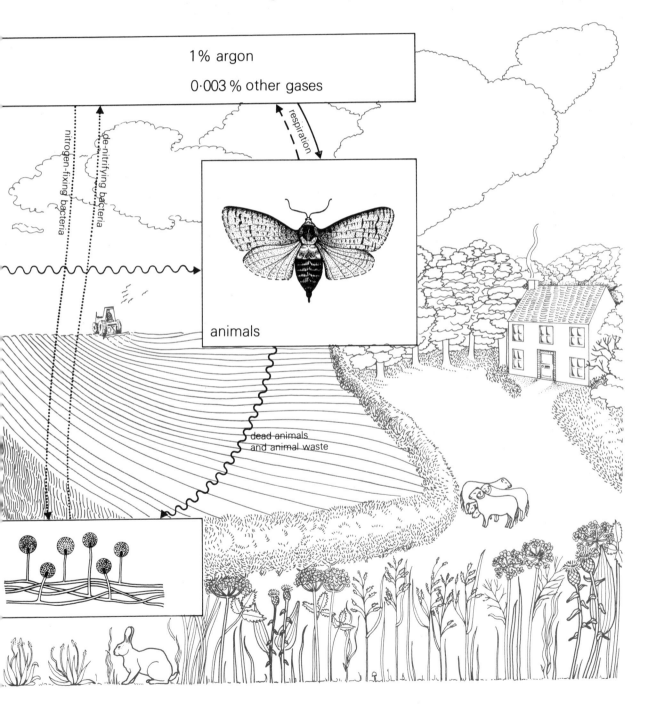

1% argon

0·003 % other gases

respiration

de-nitrifying bacteria

nitrogen-fixing bacteria

animals

dead animals and animal waste

would not be able to respire, and without respiration they would not be able to live.

For twenty-four hours a day, oxygen is being taken from the atmosphere for use in respiration (see continuous-line arrows in the chart). Oxygen is also taken when we burn fuels. To maintain the supply, this oxygen must be replaced as rapidly as it is removed. The only organisms that *give* oxygen to the atmosphere are green plants, when they photosynthesise (Chapter 2). The plants growing on land and the vast numbers of plant plankton of the oceans (Chapter 29) provide between them the right amount of oxygen to replace that used in respiration.

Respiration and burning put large quantities of carbon dioxide into the atmosphere (see dashed-line arrows on chart). In large amounts, carbon dioxide is harmful to organisms. It is a harmful condition (Chapter 1). Green plants remove carbon dioxide from the atmosphere so that, under natural conditions at least, the amount of carbon dioxide remains at the low level of 0.03 per cent. Plants are the only organisms that can do this.

The chart also shows what happens to the energy taken in by plants during photosynthesis (wavy-line arrows). This will be discussed in more detail in the next chapter. The dotted-line arrows show transfer of nitrogen gas between the atmosphere and certain soil organisms, as explained in Chapter 8.

Things to do

1 Make a chart like the one in this chapter, to show what gases enter and leave the atmosphere in your neighbourhood. In the 'boxes' you could draw pictures of organisms commonly found in your area instead of the tree, moth, and other organisms shown in the chart. You could add special 'boxes' to your chart to show local industries or other activities that add gases to the atmosphere or remove gases from it. Gases added might include harmful gases produced by industrial processes.

2 Study the chart (including any chart you have made) and then answer these questions:
(a) Can you say why a tree was chosen as an example of a plant for the chart in this chapter? Is there any other type of plant that would make a good example to choose to draw in the 'box'?
(b) Can you say why an insect was chosen as an example of an animal for the chart in this chapter?
(c) When we burn coal we obtain energy, in the form of heat and light. Explain where this energy comes from.
(d) In industrial countries, the amounts of energy required are increasing. In what ways do we get energy for factories, for transport? What effect on the atmosphere would you expect to find with an increasing use of energy from fuels?
(e) In many parts of the world, large areas of forest are being cleared to make way for new towns, roads and railways. What effect is this likely to have on the atmosphere?
(f) What are the effects on the atmosphere of a large forest fire?
(g) It has been found that the percentage of carbon dioxide in the air of an area of woodland changes slightly during the day and night. At what time of day (or night) would you expect the percentage to be highest? At what time would you expect it to be lowest? Assume that the weather is calm and windless.

31 Lifestyles

Caterpillars, cows, squirrels and snails are very different from each other, but they all have one important feature in common. They all feed on plants. To that extent, they all have the same lifestyle. They are all herbivores. This illustrates the fact that organisms of many different kinds may all share the same general lifestyle.

Some organisms do not fit neatly into our scheme of lifestyles. The omnivores, for example, are generally not specialised for eating plants or for feeding on flesh, though they can make up for this by being able to accept whatever type of food happens to be available. In a similar way, as explained in Chapter 18, the damping-off fungus first lives as a parasite, then as a decomposer. The sundew plant (Figure 8.6) photosynthesises normally, so is a true primary producer, yet it can also obtain food materials from the insects it catches and digests. To a certain extent, it belongs also among the carnivores. The living world is complicated and shows great variety – not everything fits exactly into the simple system of lifestyles that has been invented for this book.

The chart shows the way in which the organisms of each kind of lifestyle depend on each other for organic food. Only one type of organism does *not* get organic food from one of the others – which type is this? How do organisms of this type get their organic food? (Hint: see Chapter 2).

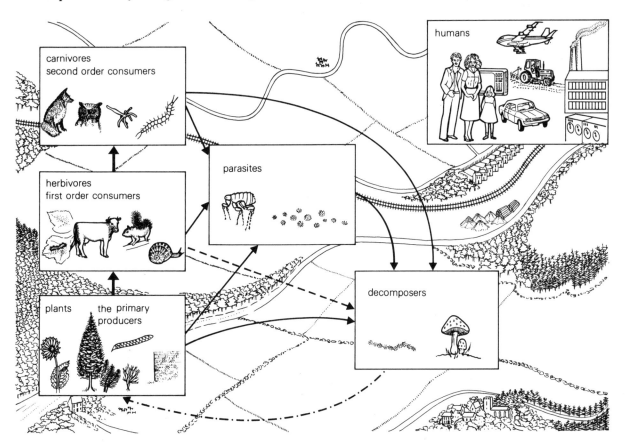

fig 31.1

Organic food provides materials for growth and for reproduction. It also provides energy. When herbivores eat plants they are getting the materials they need for building their bodies. They are getting the energy they need for moving around, for keeping warm (if they are warm-blooded) and in general for every activity of their lives. Without a supply of plants to eat, they cannot live. Much of the energy that they get from plants is lost to their surroundings, as heat. Some of the material they get from plants is passed out of their bodies as waste material, and some becomes the food of decomposers (dashed-line arrows in chart). If herbivores die without being eaten by carnivores or carrion-feeders (p. 99), decomposers feed on their remains.

Carnivores get their food by eating herbivores or other carnivores. The organic material they eat has already been transferred from the bodies of plants to the bodies of the herbivores. Now it is being incorporated into the bodies of carnivores. This is why the carnivores are sometimes called second order consumers. They consume materials that have already been consumed once before, by the herbivores.

The chart shows that the main transfer of energy and body-building materials is from plants to herbivores, and from herbivores to carnivores. There is also a transfer from carnivores to the bigger and more powerful carnivores (for example, lion, kestrel, pike, shark, eagle, wolf). These are known as *top carnivores*. At each stage of transfer along this main chain:

energy is lost to the surroundings—mainly as heat

energy and **materials** (in the form of food) are lost to the parasites

energy and **food** (in the form of waste and dead bodies) are lost to the decomposers.

Mineral salts, released by decomposition of waste and dead remains of plants and animals are returned to the soil, where they can be used again by plants (dot-and-dashed arrow in chart).

Together the five main lifestyles (plants, herbivores, carnivores, parasites and decomposers) make a self-contained system of living organisms that has survived on Earth for millions of years. Mineral materials from the soil are recycled and used over and over again. Energy is not used over and over again. It enters the system at one point (as what? where?) and leaves it at all points (as what?).

The sixth lifestyle has appeared on Earth only recently. The technological lifestyle is very different from that of the other organisms.

Things to do

1 Make a chart like the one in this chapter, but fill it with examples taken from a place in your district, such as a wood, a park or garden, a pond, a moor, a hedge or a meadow. In the 'box' for humans, put drawings to illustrate the ways in which humans affect the other plants and animals living in the area. You could include drawings to show what plants and animals (if any) are taken for food and what good or harm is done to the plants and animals by humans living in or near to the area you are studying.

2 Study the chart shown in this chapter and any similar chart you have made. Discuss the answers to these questions in class:
(a) In any given area there are usually fewer carnivores than herbivores, and there are usually very few top carnivores. Explain the reason for this.
(b) The chart shows no arrows connecting the human lifestyle with the others. What arrows can be drawn to connect us to the other members of the living world?
(c) Is our technology good for us, in general?
(d) Is our technology good for other organisms, in general?

Teacher's notes

The chapters are arranged to allow for the seasons' changes. The book may be divided between the three terms: Autumn Term, Chapters 1 – 12; Winter Term, Chapters 13 – 21; Summer Term, Chapters 22 – 31. The approach is deliberately non-quantitative, but the field studies and much of the practical work on microbiology can be made quantitative.

Chapter 1

Supplement this with new information from subsequent space flights. Harmful conditions include pollution of air, water and soil: there is scope here for discussions in class.

Chapter 2

Photosynthetic bacteria are omitted because of their minor role in the biosphere today. The demonstrations may be performed by the teacher, or one may be allocated to each of several groups of pupils. The chapter can serve as a summary to practical work on leaf anatomy and morphology, or the chapter may be studied first and practical work can follow.

Chapter 3

Simple techniques can show a lot, especially if only *small* pieces of material are examined. Wedge-shaped sections usually show good detail at the thin edge. Colour transparencies of photomicrographs are useful supplementary material.

Chapter 4

Examine scrapings from fences or bark (*Pleurococcus*), pond water, damp mud, aquarium water and living cultures. Study microscope slides, slides of colour photomicrographs, photographs from books. Certain species can be studied more completely and in more detail than in the chapter if this is required for examination purposes. Locomotion, feeding, binary fission and other activities are best demonstrated by examining cultures and by cassette films. A supply of blank tables is useful for item 4 of 'Things to do', and can be easily run off on a duplicating machine. The Vitalchrome Protokit (from Gerrard, address p. 117) includes an interesting culture of mixed protozoa, materials for vital staining and for slowing the locomotion of micro-organisms. The quantities of materials supplied with the kit are ample for examining organisms obtained from other sources too, such as those listed at the beginning of this paragraph.

Chapter 5

Practical study of seed structure and germination could precede study of this chapter. Specimens illustrating seed-dispersal mechanisms may be collected and displayed.

Chapter 6

This is not intended to provide for a detailed study of any one or more species (though this can be done if required for examination purposes) but to consider fungi in general with emphasis on their role as decomposers. Parasitic fungi are considered in Chapter 18. The Gerrard Introductory Microbiology Kit provides materials and detailed instructions for most of the practical work mentioned in this chapter (e.g. Figure 6.7 parts 4 and 5) and in Chapters 7, 9, 10 and 17. The media and broths supplied are ready-prepared and sterile. The investigations may be performed either as a demonstration by the teacher, or they may be allocated to groups of pupils, working under supervision. The Gerrard Basic Microbiology

Kit contains culture media, sterilised and prepared ready for pouring into the sterile, disposable petri dishes supplied with the kit. An autoclave is not required. The kit also contains cultures of a selection of fungi that are suitable for illustrating this chapter. Among them is *Aspergillus oryzae*, which can be grown on starch agar; subsequent flooding with iodine solution demonstrates the extra-mycelial digestion of food materials (see second paragraph of the chapter).

Chapter 7

The Home Economics Department of the school may be willing to assist with demonstrations of techniques for making bread, yoghurt and various cheeses. An outing to a local dairy product manufacturer could be timed to coincide with study of this chapter. Technology Module no. 4, *Food Technology* (NCST, address p. 117) contains many helpful suggestions for pupil activities, visual resources, hardware and texts, relevant to this chapter and especially to chapters 11 and 12.

Chapter 8

The main point of this chapter is to present bacteria as an essential and useful part of the ecosystem. The photographs provide a basis for discussion and further activities.

Chapter 9

There is not sufficient space to go into details of particular diseases; the aims of this chapter are to put bacterial disease into its biological perspective, and to introduce the parasitic lifestyle (which is dealt with in more detail in Chapter 18.). Those teachers who wish to orient their course towards human biology have scope in this chapter, and in several later chapters, to amplify the text with other materials, including film-strips and wall-charts on topics in human biology. 'Things to do', especially item 1, provides opportunity for project work on disease.

Groups could investigate topics such as: personal hygiene, food hygiene, tooth decay, houseflies; first-aid treatment of wounds and airborne infection; they could make studies of individual diseases such as whooping cough, typhoid, cholera, dysentery and tuberculosis. Viral diseases could be included. Practical demonstrations can include touching unwashed finger-tips on to sterile nutrient agar followed by incubation of the dishes after sealing with tape. Examination of the sealed dishes a few days later will demonstrate the occurrence and numbers of bacteria on the finger-tips. Parallel investigations can show the effects of washing the fingers using various procedures for washing and for drying the fingers after washing. The effectiveness (or otherwise) of the procedures can then be assessed. It is also easy to investigate the occurrence of bacteria on kitchen utensils and the working surfaces used for preparing food.

Chapter 10

Samples of water from various sources can be tested by taking a loop-ful of each and smearing it over sterile nutrient agar. Examination of the sealed dish a few days later enables counts of colonies to be made, showing the relative amount of bacterial contamination. It is important that the samples be collected in sterile containers and tested as soon as possible after collection. Study of this chapter should be linked to consideration of methods used locally for the supply and treatment of water.

Chapter 11

This chapter has strong links with historical studies, and could be supplemented by visits to museums of archaeology and local crafts. Some of the sets of slides and the overhead transparencies on *Social Biology* by Charles Brady (available from Audio-Visual Productions, see p. 117) are illustrative material for this chapter and several later chapters.

Chapter 12

Several films on food preservation and related topics are available on loan from Unilever (p. 117) for a small handling charge. Pasteurisation may be demonstrated by using milk that has stood for a day in a warm room (to *un*pasteurise it!). Then, samples are (a) left without further treatment; (b) heated for thirty minutes at 62.8°C (the legal Low Temperature Holding conditions) and then cooled rapidly; (c) brought to the boil for one minute and (d) sterilised in an autoclave for twenty minutes at 120°C in a sealed container. Samples are then inspected daily for signs of spoilage. A demonstration of the effect of heat treatment is included in the Gerrard Introductory Microbiology Kit (p. 117).

Chapter 13

Films on fibres and textiles are available on free loan from ICI (p. 117).

Chapter 14

Educoll 1 and 2 (UNICEF, p. 117) consists of eight cut-out cardboard models of dwellings from various regions of Africa. They add interest to a display of types of dwellings and show the use of a wide variety of building materials. Additional practical work could include an examination of building materials to measure (or at least place in rank-order) their properties, such as mechanical strength, waterproofness and thermal insulation. Nuffield Secondary Science Theme 7, *Using Materials*, contains many suggestions for practical activities on this and related topics. Technology Module no. 6, *Plastics* (NCST, p. 117) has some relevance to this chapter.

Chapter 15

Include discussion of items of local and national news related to the prevention and cure of pollution.

Chapter 16

Base this chapter on a study of specimens and photographs of as wide a variety of insect species as can be assembled. The film *The Rival World* is available on free loan from the Shell Film Library (p. 117) and is highly recommended. Amongst other topics, it introduces the idea of insects as vectors of disease, the subject of chapter 17. Shell Education Service (p. 117) publishes useful wall-charts (free) and photographs of insects and their life stories.

Chapter 17

The idea of metamorphosis may be mentioned briefly here, but is dealt with more fully in chapter 22. Relevant films are available from ICI and Shell (p. 117).

Chapter 19

The Gerrard microbiology kits contain demonstrations on the effectiveness of antiseptics and on the production and effectiveness of antibiotics.

Chapter 20

Practical studies of the growth and development of tulip, daffodil, crocus, carrot, potato and iris (or Solomon's seal) may be started now, and the various stages observed and noted occasionally during the following months. An excellent reference book for this topic is *The Oxford Book of Food Plants*, by Masefield *et al* (Oxford University Press). Many of the larger reference atlases include data on the food crops commonly cultivated in different geographical regions. The Unilever Film Library (p. 117) has several films on oilseeds.

Chapter 21

As in Chapter 16, cover as wide a variety of species as possible, concentrating attention on the more unusual and interesting features of each. A *few* species may then be studied in fuller detail, if required for examination purposes.

Insects of various species may be reared in the laboratory during the summer term. These include species regularly reared in school (locust, flour beetle, fruit fly), insects reared from eggs or larvae obtained from suppliers (butterfly, moth, stick insect) and any specimens found during field expeditions. If possible, aquatic species, such as dragonfly, mayfly, gnat and dytiscus, should be included. Detailed advice on handling and rearing insects is given in *Animals and Plants* (Nuffield Junior Science: Collins). Other useful information is to be found in Minibeasts (Science 5 – 13: Macdonald Educational).

Chapter 23

Supplementary work could include a survey of the area around the school, identifying and mapping the trees. One common species could then be selected for special study from spring to early summer, or through until autumn. Suggestions for practical activities, both outdoors and indoors, to accompany this chapter, Chapter 24 and Chapters 26–31 can be found in the author's *Outdoor Biology* series (Murray). Further suggestions for this chapter and Chapter 25 are in the module on *Timber Technology* (NCST, p. 117). Teachers may find accounts of the chief types of natural community found in Britain in the author's *Natural Communities* (Murray). Several publications by the Forestry Commission (p. 117) are of great interest: these include *Know your conifers, Know your broadleaves, Forestry and the town school, Starting a school forest* and many booklets dealing with animals of the forest. *Forestry and the town school* contains details of the Commission's scheme for giving young trees to schools for planting and for study. A colourful and pleasantly produced booklet on tree planting, *Trees, please*, has been published by the Blue Circle Group (p. 117) in association with the Trees Council.

Chapter 24

This introduces a theme that can form the basis of several field expeditions – which need range no further than the school garden or a nearby park. Large permanent aquaria or temporary mini-aquaria can be used to illustrate this chapter.

Chapter 26

Identification of mammals from their skulls is assisted by the key in G. B. Corbet's *The Identification of British Mammals* (British Museum, Natural History, see p. 117). This chapter could possibly be linked with a study of dental health films available from Unilever (p. 117).

Chapter 27

This should be based on field studies and include as wide a range of species as possible. A visit to a museum or (better) a zoo would help extend the range even further and provide the experience so essential as a preparation for work on classification of plants and animals in later years. Camouflage is a topic that can be introduced here; its advantages both to predator and to prey can be illustrated by many common examples.

Chapters 30 and 31

These bring together many of the points raised in earlier chapters. The charts offer scope for summarising the observations made during field studies, for discussions in class of such topics as the impact of technology on our environment and, finally, for a revision of the year's course.

Addresses

Audio-Visual Productions
15 Temple Sheen Road
London SW14 7PY

Blue Circle Information Service
The Blue Circle Group
Portland House
Stag Place
London SW1E 5BJ

British Museum (Natural History)
Cromwell Road
London SW7 5BD

Forestry Commission
231 Corstorphine Road
Edinburgh
EH12 7AT

(Details of all Forestry Commission publications are given in their *Publications List*; publications are available from the Government Bookshop, 49 High Holborn, London WC1V 6HB, or by post from PO Box 569, London SE1 9NH or from booksellers.)

T Gerrard and Co.
East Preston
Sussex

(These are the sole suppliers of the microbiology kits mentioned on p. 113)

National Centre for School Technology (NCST)
Trent Polytechnic
Burton Street
Nottingham
NG1 4BU

Shell Education Service
Shell UK Limited
Shell Mex House
Strand
London WC2R 0DX

(Refer to their latest catalogue for details.)

UNICEF Greeting Cards
84 Broomfield Road
Chelmsford
Essex CM1 1SS

Libraries which have 16 mm films available on free loan

ICI Film Library
Thames House North
Millbank
London SW1P 4QG

(Malaria, fibres and textiles, soil and many other topics.)

Shell Film Library
25 The Burroughs
Hendon
London NW4 4AT
(The spread of disease, water pollution, agricultural technology, soil, insects and many other topics; some items from this library are also available as video-cassettes.)

Unilever films can be obtained from:

National Audio-Visual Aids Library
Paxton Place
Gipsy Road
London SE27 9SR

and from:
Scottish Central Film Library
16–17 Woodside Terrace
Glasgow G3 7XM

(A small handling charge is made. Subjects include oilseeds, margarine, food technology and food hygiene, human physiology and dental health.)

Index